I0021246

Mastering Geospatial Development with QGIS 3.x
Third Edition

An in-depth guide to becoming proficient in spatial data
analysis using QGIS 3.4 and 3.6 with Python

Shammunul Islam
Simon Miles
Kurt Menke, GISP
Richard Smith Jr., GISP
Luigi Pirelli
John Van Hoesen, GISP

BIRMINGHAM - MUMBAI

Mastering Geospatial Development with QGIS 3.x
Third Edition

Copyright © 2019 Packt Publishing

All rights reserved. No part of this book may be reproduced, stored in a retrieval system, or transmitted in any form or by any means, without the prior written permission of the publisher, except in the case of brief quotations embedded in critical articles or reviews.

Every effort has been made in the preparation of this book to ensure the accuracy of the information presented. However, the information contained in this book is sold without warranty, either express or implied. Neither the authors, nor Packt Publishing or its dealers and distributors, will be held liable for any damages caused or alleged to have been caused directly or indirectly by this book.

Packt Publishing has endeavored to provide trademark information about all of the companies and products mentioned in this book by the appropriate use of capitals. However, Packt Publishing cannot guarantee the accuracy of this information.

Commissioning Editor: Richa Tripathi
Acquisition Editor: Yogesh Deokar
Content Development Editor: Unnati Guha
Technical Editor: Dinesh Chaudhary
Copy Editor: Safis Editing
Project Coordinator: Manthan Patel
Proofreader: Safis Editing
Indexer: Priyanka Dhadke
Graphics: Jisha Chirayil
Production Coordinator: Aparna Bhagat

First published: March 2015
Second edition: September 2016
Third edition: March 2019

Production reference: 1270319

Published by Packt Publishing Ltd.
Livery Place
35 Livery Street
Birmingham
B3 2PB, UK.

ISBN 978-1-78899-989-2

www.packtpub.com

`mapt.io`

Mapt is an online digital library that gives you full access to over 5,000 books and videos, as well as industry leading tools to help you plan your personal development and advance your career. For more information, please visit our website.

Why subscribe?

- Spend less time learning and more time coding with practical eBooks and Videos from over 4,000 industry professionals

- Improve your learning with Skill Plans built especially for you

- Get a free eBook or video every month

- Mapt is fully searchable

- Copy and paste, print, and bookmark content

Packt.com

Did you know that Packt offers eBook versions of every book published, with PDF and ePub files available? You can upgrade to the eBook version at `www.packt.com` and as a print book customer, you are entitled to a discount on the eBook copy. Get in touch with us at `customercare@packtpub.com` for more details.

At `www.packt.com`, you can also read a collection of free technical articles, sign up for a range of free newsletters, and receive exclusive discounts and offers on Packt books and eBooks.

Contributors

About the authors

Shammunul Islam is a consulting spatial data scientist at the Institute of Remote Sensing, Jahangirnagar University, and a senior consultant at ERI, Bangladesh. He develops applications for automating geospatial and statistical analysis in different domains such as in the fields of the environment, climate, and socio-economy. He also consults as a survey statistician and provides corporate training on data science to businesses. Shammunul holds an MA in Climate and Society from Columbia University, an MA in development studies, and a BSc in statistics.

Simon Miles is a GIS/web developer working for the Royal Borough of Windsor and Maidenhead, and has been working in the GIS industry since 2003. Both in the office and at home, he works using open source technologies. His main kit or stack includes QGIS, PostGIS, Python, GDAL, Leaflet, JavaScript, and PHP. His main focus presently is on digital delivery/transformation, chiefly through the consolidation of data, applications, and legacy systems, and replacing these with APIs and web applications. In 2013, Simon helped to establish the UK QGIS user group, which holds regional meetings two or three times a year.

Kurt Menke, **GISP**, a former Albuquerque, New Mexico, USA-based archaeologist, has an MS in geography from the University of New Mexico in 2000. He has founded Bird's Eye View to apply his expertise with GIS technology toward ecological conservation. His other focus areas are public health and education. He is an avid GIS proponent since he began writing MapServer applications in 2002. He has authored *Discover QGIS*. In 2015, Kurt was voted in as an OSGeo Charter Member. He is a FOSS4G educator and a co-author of the GeoAcademy. In 2015, he was awarded the Global Educator of the Year Team Award by GeoForAll as part of the GeoAcademy team. He authors an award-winning blog on FOSS4G technologies and their use in community health mapping.

Richard Smith Jr., GISP, is an assistant professor of geographic information science in the School of Engineering and Computing Sciences at Texas A&M University – Corpus Christi. He has a Ph.D. in geography from the University of Georgia and holds an MSc in computer science and a BSc in geographic information science from Texas A&M University-Corpus Christi. Richard actively researches in cartography, systems integration, and the use of geospatial technology for disaster response. He is an advocate for FOSS4G and building a FOSS4G curriculum. He is also one of the co-authors of the FOSS4G Academy.

Luigi Pirelli is a QGIS core developer and software analyst with a degree in computer science from Bari University. He worked for 15 years in the Satellite Ground Segment and Direct Ingestion for the European Space Agency. He is also involved in GFOSS world, contributing in QGIS, GRASS, & MapServer core, and developing and maintaining many QGIS plugins. Luigi is the founder of the OSGEO Italian local GFOSS chapter. He has taught PyQGIS, delivering training from basic to advanced levels, and supporting companies to develop their own specific QGIS plugins. He founded a local hackerspace group. Bricolabs.cc. He likes training groups on conflict resolution. Other than this book, he has also contributed to the Lonely Planet guide *Cycling Italy*.

John Van Hoesen, GISP, is an associate professor of geology and environmental studies at Green Mountain College in rural west-central Vermont, USA. He earned an MS and a Ph.D. in geology from the University of Nevada, Las Vegas, in 2000 and 2003. He is a certified GISP with a broad background in the geosciences and has used some flavor of GIS to evaluate and explore geological processes and environmental issues since 1997. John has used and taught graduate, undergraduate, and continuing education courses using some variants of FOSS GIS since 2003.

About the reviewers

Giuseppe De Marco, a Ferentino based agricultural engineer, has a Bachelor's in agriculture from the University of Pisa. He started programming at an early age. He developed deep interests in geography and GIS during his bachelor's. He then got introduced to GRASS and QGIS while working with Eris commercial products. Since QGIS 1.7.4, he has been developing plugins for it, In 2008, he partnered with his 2 colleagues and started Pienocampo (open field), a website that hosts plugins made by him which are also hosted on OGIS official repository. He likes studying geography, surveying, tree risk assessment, landscaping, bioengineering, and farm consulting. He also likes imparting knowledge on how to use QGIs and other open source software.

Chima Obi is the lead geospatial analyst at AGERPoint Inc. He has over 4 years of experience as a geospatial analyst. His specialty includes processing LIDAR data and feature extraction from raster/imagery data utilizing Python and R programing, as well as exploring other open source geospatial tools. He got his Bachelors in soil science from the Federal University of Technology, Owerri, Nigeria in 2010. He moved to the United States where he obtained his master's degree in environmental science and got a certification in Geospatial information systems in 2016.
Prior to working at AGERPoint, he worked as a geospatial analyst at West Virginia District of Highways in 2015 and 2016.

> *I would like to express my gratitude to my friends and family, and most importantly to my wife, for their wonderful encouragement and support. Also, my biggest thanks go to Packt Publishing for choosing me to be a part of this awesome book review.*

Packt is searching for authors like you

If you're interested in becoming an author for Packt, please visit `authors.packtpub.com` and apply today. We have worked with thousands of developers and tech professionals, just like you, to help them share their insight with the global tech community. You can make a general application, apply for a specific hot topic that we are recruiting an author for, or submit your own idea.

Table of Contents

Preface   1

Section 1: Introduction

Chapter 1: A Refreshing Look at QGIS   9
  Release schedules   10
  QGIS downloading and installation   10
    Installing QGIS on Windows   10
    Installing QGIS on macOS   11
    Installing QGIS on Ubuntu Linux   11
      Installing QGIS only   11
      Installing QGIS and other FOSSGIS packages   12
    QGIS on Android   12
    Installing QGIS on a Chromebook   13
    QGIS in a browser   13
  What's new?   13
  A tour of QGIS   16
    Familiarizing yourself with QGIS Desktop   16
    Loading data into QGIS Desktop   17
    Loading vector data   18
    Loading raster data   19
    Loading databases   21
    Loading web services   21
  Working with CRS   22
  Working with tables   22
    Creating table joins   24
  Editing data   26
    Snapping   27
    Styling vector data   28
    Styling raster data   30
    Blending modes   32
  Composing maps   32
  Adding functionality with plugins   33
    Custom QGIS variables   35
  Summary   36

Section 2: Getting Started

Chapter 2: Styling Raster and Vector Data   39
  Choosing and managing colors   39

Knowing color picker components   42
Changeable panels in the color picker   43
    Color ramp   43
    Color wheel   44
    Color swatches   44
    Color sampler   46
Color picker components in the Layers panel   47

Managing color ramps   49
Managing the QGIS color ramp collection   49
    Renaming a color ramp   51
    Removing a color ramp   51
    Exporting a color ramp   51
    Importing a color ramp   52
    Adding a color ramp   53
        Adding a gradient color ramp   53
        Adding a random color ramp   55
        Adding a ColorBrewer color ramp   56
        Adding a cpt-city color ramp   57
    Editing a color ramp   59
Styling singleband rasters   59
Paletted raster band rendering   59
Singleband gray raster band rendering   61
Singleband pseudocolor raster band rendering   64
    Singleband pseudocolor interpolations   67

Styling multiband rasters   68
Raster color rendering   68
Raster resampling   71
Styling vectors   72
Single-symbol vector styling   73
Categorized vector styling   76
Graduated vector styling   77
Rule-based vector styling   79
Point-displacement vector styling   82
Point cluster vector styling   84
Inverted polygons vector styling   84
Heatmap vector styling   86
2.5 D vector styling   88
Vector layer rendering   91
Layer transparency   91
Layer blending mode   91
Feature blending mode   92
Control feature rendering order   93
Using diagrams to display thematic data   94
Parameters that are common to all diagram types   95
    Diagram size parameters   95
    Diagram placement parameters   96
    Adding attributes to diagrams   98

Creating a pie chart diagram | 99
Creating a text diagram | 101
Creating a histogram chart diagram | 102
Saving, loading, and setting default styles | 103
Saving a style | 104
Loading a style | 104
Setting and restoring a default style | 104
Adding and renaming styles in the current QGIS project | 105
Summary | 105
Chapter 3: Creating Spatial Databases | 107
Fundamental database concepts | 108
Describing database tables | 108
Knowing about table relationships | 109
Reviewing the basics of the structured query language | 100
Creating a spatial database | 110
Connecting to a GeoPackage | 112
Importing data into a GeoPackage | 112
Importing a vector file | 112
Importing a layer from map canvas | 113
Working with tables | 114
Creating a new table within an existing GeoPackage | 114
Renaming a table | 115
Editing an existing table field/column | 115
Deleting a GeoPackage table | 116
Exporting tables out of a GeoPackage | 116
Creating queries and views | 117
Using select statements | 117
Creating a spatial view | 118
Dropping a spatial view | 118
Summary | 119
Chapter 4: Preparing Vector Data for Processing | 121
Merging vectors | 122
Converting vector geometries | 124
Creating polygon centroids | 124
Converting lines into polygons | 125
Converting polygons into lines | 126
Creating polygons surrounding individual points | 127
Voronoi polygons | 127
Delaunay triangulation | 130
Extracting nodes (vertices) | 132
Simplifying and densifying features | 133
Converting between multipart and singleparts features | 134
Adding geometry columns to an attribute table | 135

Using basic vector Geoprocessing Tools   135
     Spatial overlay tools   136
     Using the Clip and Difference tools   136
     Using the Intersect and Symmetrical Difference tools   140
     Overlaying polygon layers with Union   141
     Creating buffers   142
     Generating convex hulls   145
     Dissolving features   145
Defining coordinate reference systems   145
     Understanding the PROJ.4 definition format   146
     Defining a new custom coordinate reference system   147
         Setting Definitions   147
         Viewing a statistical summary of vector layers   148
Advanced field calculations   149
     Exploring the field calculator interface   149
     Writing advanced field calculations   151
         Calculating and formatting current date   151
         Calculating with geometry   152
     Operators   153
     Conditions   153
     Conditionals   153
Summary   154

Chapter 5: Preparing Raster Data for Processing   155
Merging rasters   155
About converting raster files   158
     Translating   158
     Exporting to a raster   160
         Exporting a raster to a GeoPackage   160
Clipping a raster   161
Converting rasters into vectors   163
Converting from vector to raster (rasterize)   164
Reclassifying rasters   166
Summary   169

Section 3: Diving Deeper

Chapter 6: Advanced Data Creation and Editing   173
What's new in editing?   174
     CAD-style digitizing tools   174
     Adding a circle   174
         Adding a circle from two points   175
         Adding a circle from three points   176
     Adding a rectangle   177
         Adding a rectangle from Extent   178
         Adding a rectangle from its center point   178
         Adding a rectangle from three points   178

Adding a regular polygon   178
   Adding a regular polygon from the center and from a corner   179
   Adding a regular polygon from two points   179
Vertex tool   179
Creating points from coordinate data   181
Mapping well-known text representations of geometry   185
Geocoding address-based data   188
How address Geocoding works   189
The first example – Geocoding using web services   190
The second example – Geocoding using local street network data   192
Georeferencing imagery   196
Understanding ground control points   196
Using the Georeferencer GDAL plugin   197
The first example – Georeferencing using a second dataset   202
   Getting started   202
   Entering ground control points   205
   Transformation settings   207
   Completing the operation   213
The second example – Georeferencing using a point file   214
Checking the topology of vector data   217
Installing the topology checker   217
Topological rules   218
   Rules for point features   218
   Rules for line features   218
   Rules for polygon features   219
Using the topology checker   220
Repairing topological errors via topological editing   224
Example 1 – Resolving duplicate geometries   225
Example 2 – Repairing overlaps   225
   Setting the editing parameters   226
Repairing an overlap between polygons   229
Example 3 – Repairing a gap between polygons   230
Summary   232

Chapter 7: Advanced Data Visualization   233
Using live layer effects   234
Creating beautiful effects with inverted polygon shapeburst fills   239
Creating coastal vignettes   240
Studying area mask   242
Using the 2.5D renderer   244
Creating 3D views   249
Creating an Atlas   250
Basic Atlas configuration   250
Dynamic titles   252
Dynamic legends   255
Highlighting the coverage feature   255

The power of geometry generators   258
Working with the Data Plotly plugin   262
Summary   265

Section 4: Becoming a Master

Chapter 8: The Processing Toolbox   269
   Introducing the Processing Toolbox   269
   What's new in the Processing Toolbox?   270
   Configuring the Processing Toolbox   271
   Viewing the Processing Toolbox   272
     Running algorithms in the Processing Toolbox   274
   Using the Processing Toolbox   276
   Performing raster analysis with GRASS   276
     Calculating shaded relief   278
     Calculating least-cost path   280
       Calculating slope using r.slope   281
       Reclassifying the new slope raster and the land use raster   282
       Combining reclassified slope and land use layers   284
       Calculating the cumulative cost raster using r.cost   285
       Calculating the cost path using LCP   286
     Evaluating a viewshed   288
       Clipping elevation to the boundary of the park using GDAL   289
       Calculating viewsheds for towers using r.viewshed   290
       Combining viewsheds using r.mapcalc.simple   292
   Performing analysis using SAGA   295
     Evaluating a habitat   296
       Calculating elevation ranges using the SAGA Raster calculator   296
       Clipping land use to the park boundary using Clip grid with polygon   298
       Querying land use for only surface water using the SAGA Raster calculator   298
       Finding proximity to surface water using GDAL Proximity   299
       Querying the proximity for 1,000 meters of water using the GDAL Raster
       calculator   301
       Reclassifying land use using the Reclassify grid values tool   303
       Combining raster layers using the SAGA Raster calculator   304
   Exploring hydrologic analysis with SAGA   306
     Removing pits from the DEM   306
     Deriving streams   312
     Selecting the streams   313
     Delineating the streams   316
     Calculating the upstream area above Fort Klamath   317
   Summary   323

Chapter 9: Automating Workflows with the Graphical Modeler   325
   Introducing the graphical modeler   325
     Opening the graphical modeler   326
     Configuring the modeler and naming a model   328

Working with your model   332
　Adding data inputs to your model   332
　Adding algorithms to your model   335
　Running a model   339
　Editing a model   342
　Documenting a model   344
　Saving, loading, and exporting models   345
Executing model algorithms iteratively   346
Nesting models   349
Using batch processing with models   354
Converting a model into a Python script   356
Summary   356

Chapter 10: Creating QGIS Plugins with PyQGIS and Problem Solving 357
　Webography - where to get API information and PyQGIS help 358
　　PyQGIS cookbook   358
　　API documentation   359
　　The QGIS community, mailing lists, and IRC channel   359
　　　Mailing lists   360
　　　IRC channel   360
　　　The Stack Exchange community   361
　　　Sharing your knowledge and reporting issues   361
　The Python Console   362
　　Getting sample data   363
　　My first PyQGIS code snippet   364
　My second PyQGIS code snippet - looping the layer features 364
　Exploring iface and QGis   365
　Exploring a QGIS API in the Python Console   366
　Creating a plugin structure with Plugin Builder   367
　　Installing Plugin Builder   368
　　Locating plugins   368
　　Creating my first Python plugin - plugin_first   369
　　　Setting mandatory plugin parameters   370
　　　Setting optional plugin parameters   371
　　　Generating the plugin code   372
　　　Compiling the icon resource   372
　　　Plugin file structure - where and what to customize   374
　　　　Exploring main plugin files   375
　　　Plugin Builder-generated files   376
　A simple plugin example   376
　　Adding basic logic to TestPlugin   376
　　　Modifying the layout with Qt Designer   377
　　　　Adding two pull-down menus   378
　　　Modifying GUI logic   379
　　　Modifying plugin logic   379
　　　　Classifying layers   380
　　　　Populating the combo box   381

Understanding self   381
Showing and running the dialog   382
Some improvements   382
More detail of the code   382
Setting up a debugging environment   383
What is a debugger?   384
Installing Aptana   384
Setting up PYTHONPATH   388
Starting the Pydevd server   390
Connecting QGIS to the Pydevd server   391
Connecting using the Remote Debug QGIS plugin   392
Debugging session example   392
Creating a PyDev project for TestPlugin   393
Adding breakpoints   395
Debugging in action   396
Summary   397
Chapter 11: PyQGIS Scripting   399
Where to learn Python basics   399
Tabs or spaces – make your choice!   400
How to load layers   400
How to manage rasters   401
Exploring QgsRasterLayer   402
Visualizing the layer   403
Managing vector files   404
Managing database vectors   404
Vector structure   406
The basic vector methods   406
Describing the vector structure   407
Describing the header   408
Describing the rows   409
Exploring QgsGeometry   410
Iterating over features   411
Describing iterators   413
Editing features   413
Updating the canvas and symbology   413
Editing through QgsVectorDataProvider   414
Changing a feature's geometry   415
Changing a feature's attributes   415
Deleting a feature   416
Adding a feature   417
Editing using QgsVectorLayer   418
Discovering the QgsVectorLayerEditBuffer class   419
Changing a feature's geometry   419
Changing a feature's attributes   419
Adding and removing a feature   420
Running Processing Toolbox algorithms   420

Listing all available algorithms   421
Getting algorithm information   421
Running algorithms from the console   423
Running your own processing script   425
    Creating a test Processing Toolbox script   425
    Running the script   425
Interacting with the map canvas   426
Getting the map canvas   426
Explaining Map Tools   427
Setting the current Map Tool   428
Getting point-click values   428
    Getting the current Map Tool   429
    Creating a new Map Tool   429
    Creating a map canvas event handler   429
    Creating a Map Tool event handler   430
    Setting up the new Map Tool   431
Using point-click values   432
Exploring the QgsRubberBand class   433
Summary   435

Other Books You May Enjoy   437

Index   441

Preface

Welcome to *Mastering Geospatial Development with QGIS 3.x, Third Edition*. With the release of QGIS 3, QGIS has broken new ground in enhancing the user experience with new features – this book aims to introduce you to these features. Throughout 11 chapters, you will explore QGIS 3.4 and QGIS 3.6, with particular emphasis on data processing, creation, editing, and visualization, with data stored in databases such as Spatialite, GeoPackages, and PostGIS.

Who this book is for

If you are a GIS professional, a consultant, a student, or perhaps a fast learner who wants to go beyond the basics of QGIS, then this book is for you. It will prepare you to realize the full potential of QGIS.

What this book covers

Chapter 1, *A Refreshing Look at QGIS*, covers those features that were new to QGIS in version 3.0 and those that have now been introduced at version 3.4 and version 3.6.

Chapter 2, *Styling Raster and Vector Data*, explores styling raster and vector data for display. First, color selection and color ramp management are covered. Next, single-band and multi-band raster data is styled using custom color ramps and blending modes. Then, complex vector styles and vector layer rendering are covered.

Chapter 3, *Creating Spatial Databases*, introduces the user to the use of databases that are locally stored in file directories, such as Spatialite and GeoPackages. Users will learn how to create their own local database, as well as how to add data into it from scratch, or import existing data.

Chapter 4, *Preparing Vector Data for Processing*, introduces you to the tools and functions that come as standard within QGIS for working with vector datasets.

Chapter 5, *Preparing Raster Data for Processing*, introduces you to the tools and functions that come as standard within QGIS for working with rasters.

Chapter 6, *Advanced Data Creation and Editing*, provides advanced ways to create vector data. As there is a great deal of data in tabular format, this chapter will cover mapping coordinates and addresses from tables. Next, georeferencing of imagery into a target coordinate reference system will be covered. The final portion of the chapter will cover testing topological relationships in vector data and correcting any errors via topological editing.

Chapter 7, *Advanced Data Visualization*, covers the powerful data visualization tools found only in QGIS.

Chapter 8, *The Processing Toolbox*, begins with an explanation and exploration of the QGIS Processing Toolbox. Various algorithms and tools, available in the toolbox, will be used to complete common spatial analyses and geoprocessing tasks for both raster and vector formats. To illustrate how these processing tools might be applied to real-world questions, two hypothetical scenarios are illustrated, relying heavily on the GRASS and SAGA tools.

Chapter 9, *Automating Workflows with the Graphical Modeler*, covers the purpose and use of the graphical modeler to automate analysis workflows. In this chapter, you will develop an automated tool/model that can be added to the Processing Toolbox.

Chapter 10, *Creating QGIS Plugins with PyQGIS and Problem Solving*, covers the foundational information required to create a Python plugin for QGIS. Information about the API and PyQGIS help will be covered first, followed by an introduction to the iface and QGis classes. Next, the steps required to create and structure a plugin will be covered. The chapter will be wrapped up after providing you with information on creating graphical user interfaces and setting up debugging environments to debug code easily.

Chapter 11, *PyQGIS Scripting*, covers topics on integrating Python analysis scripts with QGIS outside of the Processing Toolbox. Layer loading and management are first covered, followed by an exploration of the vector data structure. Next, the programmatic launching of other tools and external programs are covered. Lastly, the QGIS map canvas is covered with respect to how a script can interact with the map canvas and layers within.

To get the most out of this book

You should have good knowledge of Python and QGIS.

Download the example code files

You can download the example code files for this book from your account at `www.packt.com`. If you purchased this book elsewhere, you can visit `www.packt.com/support` and register to have the files emailed directly to you.

You can download the code files by following these steps:

1. Log in or register at `www.packt.com`.
2. Select the **SUPPORT** tab.
3. Click on **Code Downloads & Errata**.
4. Enter the name of the book in the **Search** box and follow the onscreen instructions.

Once the file is downloaded, please make sure that you unzip or extract the folder using the latest version of:

- WinRAR/7-Zip for Windows
- Zipeg/iZip/UnRarX for Mac
- 7-Zip/PeaZip for Linux

The code bundle for the book is also hosted on GitHub at `https://github.com/PacktPublishing/Mastering-Geospatial-Development-with-QGIS-3.x`. In case there's an update to the code, it will be updated on the existing GitHub repository.

We also have other code bundles from our rich catalog of books and videos available at `https://github.com/PacktPublishing/`. Check them out!

Also, you can download data for exercises at `https://qgis.org/downloads/data/`.

Download the color images

We also provide a PDF file that has color images of the screenshots/diagrams used in this book. You can download it here: `http://www.packtpub.com/sites/default/files/downloads/9781788999892_ColorImages.pdf`.

Conventions used

There are a number of text conventions used throughout this book.

`CodeInText`: Indicates code words in text, database table names, folder names, filenames, file extensions, pathnames, dummy URLs, user input, and Twitter handles. Here is an example: "Mount the downloaded `WebStorm-10*.dmg` disk image file as another disk in your system."

A block of code is set as follows:

```
SELECT distname as road_name, roadnumber, geom FROM SU_Road WHERE
classifica LIKE 'A Road%' AND roadnumber = 'A25'
```

Any command-line input or output is written as follows:

```
sudo apt-get update
sudo apt-get install qgis python-qgis
```

Bold: Indicates a new term, an important word, or words that you see onscreen. For example, words in menus or dialog boxes appear in the text like this. Here is an example: "Select **System info** from the **Administration** panel."

Warnings or important notes appear like this.

Tips and tricks appear like this.

Get in touch

Feedback from our readers is always welcome.

General feedback: If you have questions about any aspect of this book, mention the book title in the subject of your message and email us at customercare@packtpub.com.

Errata: Although we have taken every care to ensure the accuracy of our content, mistakes do happen. If you have found a mistake in this book, we would be grateful if you would report this to us. Please visit www.packt.com/submit-errata, selecting your book, clicking on the Errata Submission Form link, and entering the details.

Piracy: If you come across any illegal copies of our works in any form on the Internet, we would be grateful if you would provide us with the location address or website name. Please contact us at copyright@packt.com with a link to the material.

If you are interested in becoming an author: If there is a topic that you have expertise in and you are interested in either writing or contributing to a book, please visit authors.packtpub.com.

Reviews

Please leave a review. Once you have read and used this book, why not leave a review on the site that you purchased it from? Potential readers can then see and use your unbiased opinion to make purchase decisions, we at Packt can understand what you think about our products, and our authors can see your feedback on their book. Thank you!

For more information about Packt, please visit packt.com.

Section 1: Introduction

QGIS 3 is the first major new release of QGIS in several years. This section reviews what major changes will be treated as assumed knowledge for the remainder of the book.

This section contains the following chapter:

- Chapter 1, *A Refreshing Look at QGIS*

A Refreshing Look at QGIS

1

In this chapter, we will review the basic functionality of QGIS and explore some of the new features of versions starting from 3.4. If you need a refresher on QGIS or a quickstart guide to QGIS, you should read this chapter. The topics we will cover in this chapter are as follows:

- Downloading QGIS and its installation
- The QGIS graphical user interface
- Loading data
- Working with **coordinate reference systems (CRS)**
- Working with tables
- Editing data
- Styling data
- Composing a map
- Finding and installing plugins

QGIS is a volunteer-led development project licensed under the GNU General Public License, and was started by Gary Sherman in 2002. The project was incubated with the **Open Source Geospatial Foundation (OSGeo)** in 2007, with QGIS 1.0 being released in 2009. The continued development of QGIS is supported by an active and vibrant community from around the world. Many people assume that they can only help with the development of QGIS if they can perform computer programming, but this is false! QGIS has many community members that write documentation, test the program for bugs, translate documents, answer forum questions, and provide financial support. QGIS user groups exist as well, aiming to bring people together to share experiences of QGIS. It is easy to get involved, and the authors encourage you to consider contributing. Learn about how to get involved at http://qgis.org/en/site/getinvolved/.

Release schedules

Currently, a new version of QGIS is released around every four months. The version released each spring is designated as a **long-term release** (**LTR**). This means it will be supported for one calendar year. Each quarter, a new stable version is released and bug fixes applied to the LTR. The LTR is recommended for production environments since it has a slower release cycle. At the time of writing, QGIS 3.4.5 is the current LTR, with version 3.6 available for download.

 It is possible to have multiple versions of QGIS installed on a single machine.

QGIS downloading and installation

QGIS can be installed on Windows, macOS, Unix, Linux, and Android operating systems, making it a very flexible software package. Both the binary installers and source code can be downloaded from download.qgis.org. In this section, we will briefly cover how to install QGIS on Windows, macOS, and Ubuntu Linux. For the most up-to-date installation instructions, refer to the QGIS website.

Installing QGIS on Windows

On the QGIS website, download the correct version of QGIS for your Windows machine. Downloadable executable files come in 32-bit and 64-bit flavors. We recommend downloading the latest release as advertised on the QGIS website download page.

For Windows, there are two installation options, which are as follows:

- **QGIS Standalone Installer**: The standalone installer installs the binary version of QGIS and the **Geographic Resource Analysis Support System** (**GRASS**) using a standard Windows installation tool. You should choose this option if you want an easy installation experience of QGIS. This is also the ideal installation method for organizations or businesses wishing to deploy QGIS to multiple users.

- **OSGeo4W Network Installer**: This provides you with the opportunity to download either the binary or source code version of QGIS, as well as experimental releases of QGIS. Additionally, the OSGeo4W installer allows you to install other open source tools and their dependencies.

If you are unsure on how to install QGIS, the QGIS website contains details on how to do this.

Installing QGIS on macOS

The QGIS website provides download mac installer packages for macOS, that contain readme files for the installation of QGIS. Documentation and support for QGIS on macOS has greatly improved, and while it is beyond the scope of this chapter to detail the install process and document any possible issues that might incur during installation, the GIS Stack Exchange (`https://gis.stackexchange.com/`) is a great refresh resource.

Installing QGIS on Ubuntu Linux

There are two options when installing QGIS on Ubuntu: installing QGIS only, or installing QGIS as well as other **Free and Open Source Software for Geographical Information Systems** (**FOSSGIS**) packages. Either of these methods requires the use of the command line, `sudo` rights, and the `apt-get` package manager.

Installing QGIS only

Depending on whether you want to install a stable release or an experimental release, you will need to add the appropriate repository to the `/etc/apt/sources.list` file.

With `sudo` access, edit `/etc/apt/sources.list` and add the following line to install the current stable release or the current release source code, respectively:

```
deb        http://qgis.org/debian trusty main
deb-src    http://qgis.org.debian trusty main
```

Depending on the release version of Ubuntu you are using, you will need to specify the release name as `trusty`, `saucy`, or `precise`. For the latest list of QGIS releases for Ubuntu versions, visit `download.qgis.org`.

With the appropriate repository added, you can proceed with the QGIS installation by running the following commands:

```
sudo apt-get update
sudo apt-get install qgis python-qgis
```

To install the GRASS plugin (recommended), install the optional package by running this command:

```
sudo apt-get install qgis-plugin-grass
```

Installing QGIS and other FOSSGIS packages

The `ubuntugis` project installs QGIS and other FOSSGIS packages, such as GRASS on Ubuntu. To install the `ubuntugis` package, remove the `http://qgis.org/debian` lines from the `/etc/apt/sources.list` file and run the following commands:

```
sudo apt-get install python-software-properties
sudo add-apt-repository ppa:ubuntugis/ubuntugis-unstable
sudo apt-get update
sudo apt-get install qgis python-qgis qgis-plugin-grass
```

QGIS on Android

At the time of writing, there is really only one Android QGIS application available on Google Play, called **QField**, developed by OPENGIS.ch from Switzerland. Users of QGIS can save their projects to their Android device and then reopen them in QField. The application comes with Geolocation capabilities, data editing, and photos captured, to name just a few of its features.

 Further information on QField can be found at `http://www.qfield.org/docs/index.html`.

While QField may currently (at the time of writing) be one of the main Android QGIS applications available, the underlying architecture of QGIS, namely QT5 and Python3, mean that development and deployment of user-created Android applications is not far away, thanks to crowdfunding opportunities organized by such companies as Lutra Consultancy and North Road.

Installing QGIS on a Chromebook

It is possible to install the Ubuntu Linux operating system on a Chromebook via Crouton (`https://chromebook.guide/crouton/`). This essentially creates a dual boot environment allowing you to switch between the Chrome OS and Ubuntu very quickly with some keyboard strokes. Once Crouton is set up, QGIS can be installed via the preceding Ubuntu QGIS install processes.

 While Chromebooks are relatively cheap, they are weak, if not lacking, in hardware such as graphic cards. Therefore, while it is possible to install QGIS on a Chromebook, we would suggest that doing so is for hobbyists looking to explore/test the capabilities of this union and that it would not be suitable for production environments.

QGIS in a browser

Using the RollApp website (`https://www.rollapp.com`), you can access QGIS, plus a range of other open source software in your web browser. This means that for Chromebook and iPad (for example), users can access QGIS without physically installing any software or making system configuration changes. **RollApp** is designed around using cloud base storage, such as Google Drive, Dropbox, and Box. This makes sharing data and projects very easy.

What's new?

With the release of version 3 of QGIS comes a host of new features and tools—far too many to document here, but if you are interested, take a look at `http://changelog.qgis.org/en/qgis/version/3.4-LTR`. For the new addition to QGIS 3.6, have a look at this `http://changelog.qgis.org/en/qgis/version/3.6.0/`. The biggest shift in QGIS 3.x is the migration from Python 2 to 3, which has extended the functionality to run scripts, processes, and tasks. Further to this, the underlying application used to build the QGIS interface, called Qt as of version 3.x, runs off Qt5.

To highlight perhaps just a few of these new features, note the following:

- **Data Source Manager**: The **Data Source Manager** (accessed via 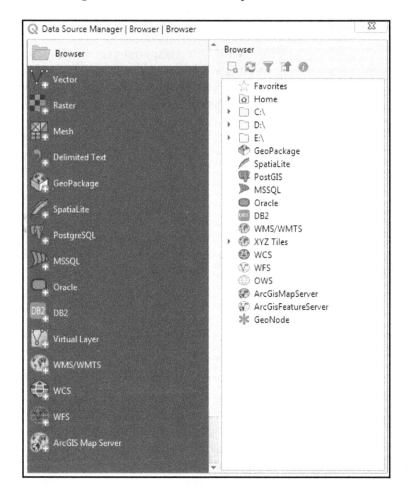 or *Ctrl + L*) is perhaps the most notable visual change of 3.4, as it is presented as the default tool to accessing data to load onto the Map Canvas:

- **Opening layers over HTTP(s), FTP, and cloud storage**: Sharing and accessing data through QGIS has gotten a whole lot easier with 3.x series. No longer does data need to be stored on a local or networked drive; instead, data can be accessed through HTTP(s), FTP, or via cloud storage providers such as Google Drive. There are now really no limits as to how QGIS can fetch, receive, and display GIS data.

- **Filters**: In previous versions of QGIS, it has been difficult to know which layers have had filters applied to them. In version 3.4, this has been remedied with a filter icon being placed on all layers that have filters applied. A user can click on the filter icon to gain quick access to the query builder.

- **Import geotagged photos**: Photos that have been geotagged (which is commonly set as a default function on most modern smartphones) can now be uploaded onto the Map Canvas and represented as a point value. The metadata associated with the photo can also be accessed.

- **Save project to Postgres**: This is a truly remarkable step forward for QGIS. The ability to save a QGIS project to a database table adds a new level of security, but also accessibility, to QGIS. It is now possible to simply send someone postgres connection details and, once added into QGIS, this will allow the user to access permission-based projects and layers (tables) stored on the database. This means that no data, no projects, and indeed, no physical files need to be handled by a user to access QGIS data.

- **Locator Search Bar**: Located in the bottom left, below the panels, is the **Locator Search** bar that is a quick finder for processes, tools, and can even be configured to search layers.

- **3D and mesh**: There has been significant investment in 3D in version 3. 3D views of data can now be seen in its own **Map View**, which runs in parallel to the main Map Canvas. Mesh data can now be loaded into QGIS and viewed in 3D.

- **Print Layouts**: Other than a change of name from Print Compose to Print Layouts, QGIS 3.4 brings some new elements to the Print window, such as the ability to add 3D views. There have been changes to the **Item Properties** to increase overall functionality.

- **Identity tool for mesh layers:** In QGIS 3.6, we can use **Identify** tool with mesh layer. We can see the value of both scalar and vector component

To have a look at the features of QGIS 3.6, please visit this link: `https://qgis.org/en/site/forusers/visualchangelog36/index.html`. For a look at the new editions in QGIS 3.4, please refer to this link: `https://qgis.org/en/site/forusers/visualchangelog34/`. If you visit any of these pages, you will find link for new features for all the versions of QGIS.

A tour of QGIS

QGIS is composed of two programs: **QGIS Desktop** and **QGIS Browser**. Desktop is used for managing, displaying, analyzing, and styling data. Browser is used for managing and previewing data. This section will give you a brief tour of the graphical user interface components of both QGIS Desktop and QGIS Browser.

Familiarizing yourself with QGIS Desktop

The QGIS interface is divided into four interface types: **Menu Bar**, **Toolbars**, **Panel**, and **Map Canvas**. The following screenshot shows QGIS Desktop with all four interface types displayed:

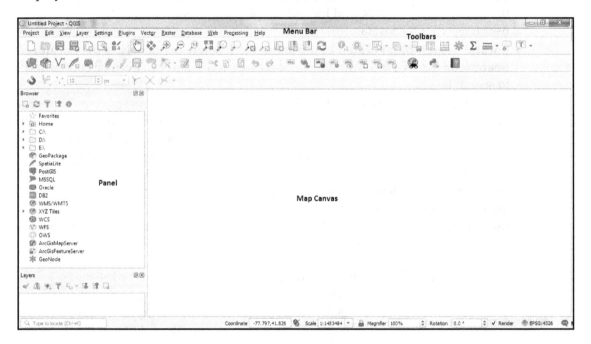

The **Map Canvas** shows the styled data added to the QGIS project. The **Menu Bar**, displayed across the top, provides access to most of the QGIS functions. The **Toolbars** provide quick access to QGIS core functionality. The **Toolbars** can be arranged to either float independently or dock at the top, bottom, left, or right sides of the application—equally, toolbar ideas can be removed from the screen via **Settings** | **Interface Customization**.

If you are going to use the Interface Customization tool to remove icons, make a backup file first of the current **UI** (**User Interface**) of QGIS. When the **Enable Customization** button is checked, click on the **Save** button and save the .ini file to a safe place. Then, make your changes but remove icons and so on. If you ever what QGIS looking as it did, prior to the removal of icons that you made, use the **Load from file** option to reinstall the original backup .ini file that you created.

Note that if you change the UI via the Interface Customization tool, then you'll need to close QGSIS before you see the changes take effect.

The panels, such as **Browser** and **Layers**, provide a variety of functionality and can be arranged to either float independently or dock above, below, right, or left of the map display or side by side as tabs.

QGIS Desktop offers a number of customization options. You can toggle the visibility of toolbars by navigating to **View** | **Toolbars**, or by right-clicking on the **Menu Bar** and then enabling the **Toolbars** button, which will open a context menu allowing you to toggle the toolbar and panel visibility. You can assign shortcut keys to operations by navigating to **Settings** | **Configure shortcuts**. You can also change application options, such as interface language and rendering options, by navigating to **Settings** | **Options**.

Loading data into QGIS Desktop

One strength of QGIS is its ability to load a large number of data types. In this section, we will cover loading various types of data into QGIS Desktop.

In general, data can be loaded in a number of ways. The main way, which will be covered in detail in this section, is to use the **Data Source Manager**.

Alternatively, using the **Browser** panel, navigate to the data you wish to load, and then either drag the data onto the **Map Canvas** or right-mouse click on it and choose **Add Selected Layer(s) to canvas**.

Under **Settings** | **Add Layer** from the main **Menu**.

Also, you can drag and drop data/files from your native operating systems file manager. Even ZIP files that contain .shp files can be added to the Map Canvas in this way.

Loading vector data

To load vector files using the Data Source Manager, click on the **Open Data Source**

Manager icon , or press *Ctrl + L*. From the list of layer types on the left, choose **Vector**. You now have the option to load a **Vector** file for a number of **Source type** options:

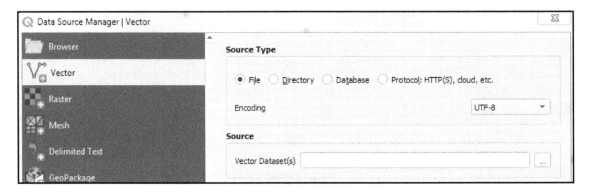

The source type contains four options: **File**, **Directory**, **Database**, and **Protocol: HTTP(S), cloud, etc.** When you choose a source type, the source interface will change to display the appropriate options. Let's take a moment to discuss what types of data these four source types can load:

> There is a big push to move users away from using ESRI shapefiles, replacing it instead with **GeoPackages** (more will be discussed about these later). QGIS, in the first instance, will prompt you to save to this format over `.shp`.

- **File**: This can load flat files that are stored on disk. The commonly used flat file types are as follows:
    - GeoPackage (`.gpkg`)
    - ESRI shapefile (`.shp`)
    - GeoJSON (`.geojson`)
    - Geography Markup Language (`.gml`)
    - AutoCAD DXF (`.dxf`)
    - Comma-separated values (`.csv`)
    - GPS eXchange Format (`.gpx`)
    - Keyhole Markup Language (`.kml`)
    - SQLite/SpatiaLite (`.sqlite`/`.db`)

- **Directory**: This can load data stored on disk that is encased in a directory. The commonly used directory types are as follows:
    - U.S. Census TIGER/Line
    - Arc/Info Binary Coverage
- **Database**: This can load databases that are stored on disk or those available through service connections. The commonly used database types are as follows:
    - ODBC
    - ESRI Personal GeoDatabase
    - MSSQL
    - MySQL
    - PostgreSQL
- **Protocol**: This can load protocols that are available at a specific URL. In QGIS 3.4, these protocols have been extended to include calls to **HTTP/HTTPS/FTP**, **AWS S3**, and **Google Cloud Storage**, adding to the previous **GeoJSON** and **CouchDB** options:

Loading raster data

To load raster files using the Data Source Manager, click on the **Open Data Source Manager** icon , or press *Ctrl + L*. From the list of layer types on the left, choose **Raster**.

You now have the option to load a raster file for **File** or **Protocol: HTTP(S), cloud, etc.** (much like **Vector**):

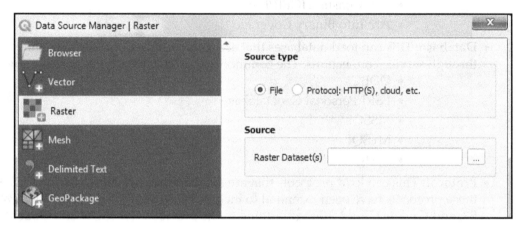

The commonly used raster types supported by GDAL are as follows:

- ArcInfo ASCII Grid (`.asc`)
- Erdas Imagine (`.img`)
- GeoTIFF (`.tif`/`.tiff`)
- JPEG/JPEG-2000 (`.jpg` or `.jpeg`/`.jp2` or `.j2k`)
- Portable Network Graphics (`.png`)
- RasterLite (`.sqlite`)
- USGS Optional ASCII DEM (`.dem`)

 The **Geospatial Data Abstraction Library** (**GDAL**) is a free and open source library that translates and processes vector and raster geospatial data formats. QGIS, as well as many other programs, uses GDAL to handle many geospatial data processing tasks.

You may see references to OGR or GDAL/OGR as you work with QGIS and GDAL. **OGR Simple Features Library** references the vector processing parts of GDAL. OGR is not really a standalone project, as it is part of the GDAL code now; however, for historical reasons, OGR is still used. More information about GDAL and OGR can be found at `http://gdal.org`. GDAL is an OSGeo (`http://osgeo.org`) project.

Loading databases

QGIS supports PostGIS, SpatiaLite, Microsoft SQL Server, and Oracle databases. Regardless of the type of database you wish to load, the loading sequence is very similar. Therefore, instead of covering specific examples, the general sequence will be covered.

To create a new database connection using the Data Source Manager, perform the following:

1. Click on the **Open Data Source Manager** icon or press *Ctrl + L*
2. On the left, choose the database you want to connect to, and, in the connection window, click on the **New** button
3. Add all the connection details required

To add a layer once a connection has been established, perform the following:

1. Click on the **Open Data Source Manager** icon
2. Click on **Connect**

When connected successfully, you will be presented with a list of layers that can be added to the Map Canvas. An alternative and perhaps better method for loading data on the **Map Canvas** is to use the Browser panel once you have created the new database connection. Using the Browser panel gives the user a cleaner overview of the layers and tables that can be added to the **Map Canvas**.

> Note that the window will look the same for any database you choose, except for the window name.

Loading web services

QGIS supports the loading of OGC-compliant web services such as WMS/WMTS, WCS, and WFS. Loading a web service is similar to loading a database service in that you must first set up the connection to the service, and then connect to the service to choose which layers to add to the Map Canvas.

As an example, to add a WMS service, click on the **Open Data Source Manager** icon or press *Ctrl + L*, and, on the left-hand side of the window, click on **WMS/WMTS**.

Click on the **New** button and give the new connection details a name (this is a free text friend, so the name can be anything you wish) and then add in the URL of the WMS/WMTS. Once completed, click on **OK**.

As with viewing data from databases in the previous section, to view the data available via the connection that you have established via the Data Source Manager, it is best to access the data via the browser panel.

Working with CRS

When working with spatial data, it is important that a CRS is assigned to the data and the QGIS project. To view the CRS for the QGIS project, click on **Properties** under **Project**, and choose the **CRS** tab.

It is recommended that all data added to a QGIS project be projected into the same CRS as the QGIS project. However, if this is not possible or convenient, QGIS can project layers on the fly to the project's CRS.

 If you want to quickly search for a CRS, you can enter the EPSG code to quickly filter through the CRS list. An EPSG code refers to a specific CRS stored in the EPSG **Geodetic Parameter Dataset** online registry that contains numerous global, regional, and local CRSs. An example of a commonly used EPSG code is 4326, which refers to WGS 84. The EPSG online registry is available at http://www.epsg-registry.org/.

To view the CRS for a layer, perform the following steps:

1. Open the layer's properties by either navigating to **Layer | Properties**, or by right-clicking on the layer in the **Layers** panel
2. Choose **Source** from the context menu
3. If the layer's CRS is not set or is incorrect, click on the globe at the end of the drop-down menu, and select the correct CRS

Working with tables

There are two types of tables you can work with in QGIS: attribute tables and standalone tables. Whether they are from a database or associated with a shapefile or a flat file, they are all treated the same. Standalone tables can be added by clicking on **Layer | Add Layer | Add Vector Layer**.

QGIS supports the table formats supported by OGR along with database tables. Tables are treated like any other GIS layer; they simply have no geometry. Both types of tables can be opened within Desktop by selecting the layer/table in the **Layers** panel, and then by either clicking on **Open Attribute Table** under **Layer**, or by right-clicking on the data layer, and choosing **Open Attribute Table** from the context menu. They can also be previewed in QGIS Browser by choosing the **Attributes** tab.

The table opens in a new window that displays the number of table rows and selected records in the title bar. Beneath the title bar is a series of buttons that allow you to toggle between editing, managing selections, and adding and deleting columns. Most of the window is filled with the table body. The table can be sorted by clicking on the column names. An arrow will appear in the column header, indicating either an ascending or a descending sort. Rows can be selected by clicking on the row number on the left-hand side. In the lower-left corner is a **Tables** menu, which allows you to manage what portions of the table should be displayed. You can choose **Show All Features** (default setting), **Show Selected Features**, **Show Features Visible on Map** (only available when you view an attribute table), **Show Edited and New Features**, create column filters, and advanced filters (expression). The lower-right corner has a toggle between the default table view and a forms view of the table.

An attribute table is shown in the following screenshot, with parts of the table window identified:

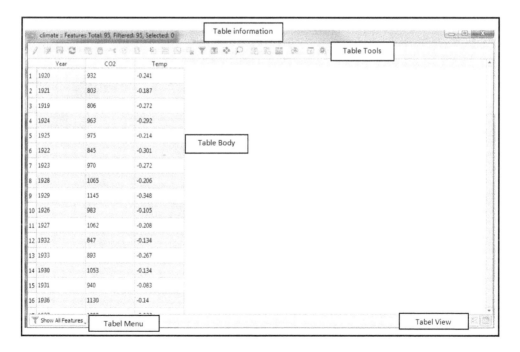

Attribute tables are associated with the features of a GIS layer. Typically, one record in the attribute table corresponds to one feature in the GIS layer. Standalone tables are not associated with GIS data layers. However, they may have data of a spatial nature from which a spatial data layer can be generated (for more information, see `Chapter 6`, *Advanced Data Creation and Editing*). They may also contain data that you wish to join to an existing attribute table with a table join, which we will cover in the next section.

Creating table joins

Let's say that you need to make a map of the total population by county. However, the counties' GIS layers do not have population as an attribute. Instead, this data is contained in an Excel spreadsheet. It is possible to join additional tabular data to an existing attribute table.

There are two requirements, which are as follows:

- The two tables need to share fields with attributes to match for joining
- There needs to be a cardinality of one-to-one or many-to-one between the attribute table and the standalone table

To create a join, load both the GIS layer and the standalone table into QGIS Desktop. QGIS will accept a variety of standalone table file formats including Excel spreadsheets (`.xls` and `.xlsx`), dBase (`.dbf`) files, and comma-separated value (`.csv`) files. You can load this tabular data using **Layer** | **Add Layer** | **Add Vector Layer** and setting the file type filter to **All files (*) (*.*)** as shown in the following screenshot:

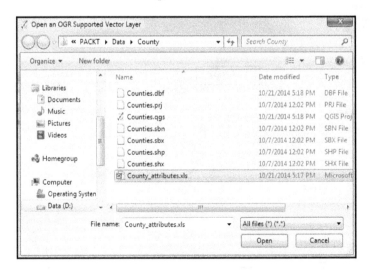

Once the data is loaded, a join can be completed by following these steps:

1. Select the GIS layer in the **Layers** panel that will receive the new data from the join.
2. Navigate to **Layer | Properties** and choose the **Joins** tab.
3. Click on the **Add join** button (the one with a green plus sign).
4. Choose the **Join layer**, **Join field**, and **Target field** values. The **Join layer** and **Join field** values represent the standalone table. The **Target field** value is the column in the attribute table on which the join will be based.

> Although, in this example, the join field and the target field have the same name, this is not a requirement. The two fields merely need to hold the same unique ID.

5. At this point, you can choose **Cache join layer in virtual memory**, **Create attribute index on join field**, or **Choose which fields are joined**. The last option allows you to choose which fields from the join layer to append to the attribute table. A new feature also allows you to set a **Custom field name prefix**. At this point, the **Add vector join** window will look like the following screenshot.
6. Once created, the join will be listed on the **Joins** tab. The extra attribute columns from the **Join layer** will be appended to the attribute table, where the value in the **Join field** matches the value in the **Target field**.
7. The additional data from the join can be used to query the data and style the data.
8. Joins can be modified by selecting the join and clicking the pencil edit button. They can be removed by clicking on the remove join button (the one with a red minus sign).

> Joins only exist in virtual memory within the QGIS Desktop document. To preserve the join outside the map document, click on **Save as...** under **Layer** and save a new copy of the layer. The new layer will include the attributes appended via the join.

Editing data

Vector data layers can be edited within QGIS Desktop. Editing allows you to add, delete, and modify features in vector datasets. The first step is to put the dataset into edit mode. Select the layer in the **Layers** panel and click on **Toggle Editing** under **Layer**. Alternatively, you can right-click on a layer in the **Layers** panel and choose **Toggle Editing** from the context menu. Multiple layers can be edited at a time. The layer currently being edited is the one selected in the **Layers** panel. Once you are in edit mode, the digitizing toolbar (shown in the following screenshot) can be used to add, delete, and modify features:

From left to right, the tools in the digitizing toolbar are as follows:

- The **Current Edits** tool allows you to manage your editing session. Here, you can save and roll back edits for one or more selected layers.
- The **Toggle Editing** tool provides an additional means to begin or end an editing session for a selected layer.
- The **Save Layer Edits** tool allows you to save edits for the selected layer(s) during an editing session.
- The **Add Features** tool will change to the appropriate geometry depending on whether a point, line, or polygon layer is selected. Points and vertices of lines and polygons are created by clicking on the **Map Canvas**. To complete a line or polygon feature, right-click. After adding a feature, you will be prompted to enter the attributes.
- The **Vertex Tool** lets you move individual feature vertices. Click on a feature once with the tool to select it, and the vertices will change into red boxes. Click again on an individual vertex to select it. The selected vertex will turn into a dark-blue box. Now, the vertex can be moved to the desired location. Additionally, edges between vertices can be selected and moved. To add vertices to a feature, simply double-click on the edge where you want the vertex to be added. Selected vertices can be deleted by pressing the *Delete* key on the keyboard.

- **Modify the attributes of all selected features simultaneously**, which allows you to set a value for all the selected features:

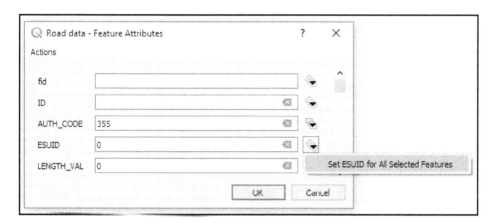

- Features can be deleted, cut, copied, and pasted using the **Delete Selected, Cut Features**, **Copy Features**, and **Paste Features** tools.

Snapping

As of version 3 of QGIS, how snapping is carried out has changed. Snapping is a specified distance (tolerance) within which vertices of one feature will automatically align with vertices of another feature. Once enabled, users can set which layers are to be used to snap against; if vertex, vertex and segments, or just segments, are to be snapped against; and the distance from which snapping will occur. Additional topological and tracing functions have been added to further enhance the snapping toolset:

Styling vector data

When you load spatial data layers into QGIS Desktop, they are styled with a random single symbol rendering. To change this, right-mouse click on the layer and choose properties, and, in the context menu, click on **Symbology**.

There are several rendering choices available from the menu in the top-left corner, which are as follows:

- **Single Symbol**: This is the default rendering in which one symbol is applied to all the features in a layer.
- **Categorized**: This allows you to choose a categorical attribute field to style the layer. Choose the field and click on **Classify** and QGIS will apply a different symbol to each unique value in the field. You can also use the **Set column expression** button to enhance the styling with a SQL expression.
- **Graduated**: This allows you to classify the data by a numeric field attribute into discrete categories. You can specify the parameters of the classification (classification type and number of classes) and use the **Set column expression** button to enhance the styling with a SQL expression.
- **Rule-based**: This is used to create custom, rule-based styling. Rules will be based on SQL expressions.
- **Point Displacement**: If you have a point layer with stacked points, this option can be used to displace the points so that they are all visible.
- **Point Cluster:** This option groups clusters of points together at a specified distance. This will be covered more in later chapters.
- **Inverted Polygons**: This is a new renderer that allows a feature polygon to be converted into a mask. For example, a city boundary polygon that is used with this renderer would become a mask around the city. It also allows the use of **Categorized**, **Graduated**, and **Rule-based** renderers and SQL expressions.
- **Heatmap**: This allows you to create a heat map rendering of point data based on an attribute.
- **2.5D**: This allows you to extrude data into a 2.5D space. A common use case is to extrude building footprints upward, creating an almost 3D rendering.

The following screenshot shows the symbology properties available for a vector line layer:

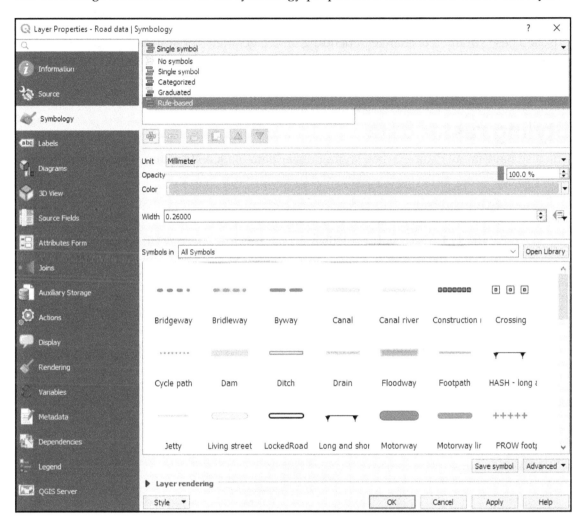

For a given symbol, you can work with the first level, which gives you the ability to change the transparency and color. You can also click on the second level, which gives you control over parameters such as fill, border, fill style, border style, join style, border width, and X/Y offsets. These parameters change depending on the geometry of your layer. You can also use this hierarchy to build symbol layers, which are styles built from several symbols that are combined vertically.

Styling raster data

You also have many choices when styling raster data in QGIS Desktop. There is a different choice of renderers for raster datasets, which are as follows:

- **Multiband color**: This is for rasters with multiple bands. It allows you to choose the band combination that you prefer.
- **Paletted / Unique values**: This is for singleband rasters with an included color table. It is likely that it will be chosen by QGIS automatically, if this is the case.
- **Singleband gray**: This allows a singleband raster or a single band of a multiband raster to be styled with either a black-to-white or white-to-black color ramp. You can control contrast enhancement and how minimum and maximum values are determined.
- **Singleband pseudocolor**: This allows a singleband raster to be styled with a variety of color ramps and classification schemes.
- **Hillshade**: Used mainly against Digital Terrain Data, this function allows to you add hillshade relief to the layer by changing the angles/direction of sunlight.

The following is a screenshot of the **Symbology** tab of a raster file's **Layer Properties**, showing where the aforementioned style choices are located:

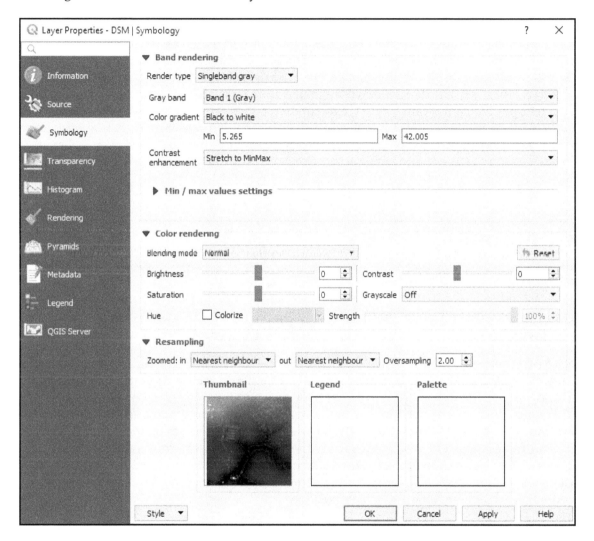

Blending modes

The blending modes allow for more sophisticated rendering between GIS layers. Historically, these tools have only been available in graphics programs and they are a fairly new addition to QGIS. Previously, only layer transparency could be controlled. There are now 13 different blending modes that are available: **Normal, Lighten, Screen, Dodge, Addition, Darken, Multiply, Burn, Overlay, Soft light, Hard light, Difference**, and **Subtract**. These are much more powerful than simple layer transparency, which can be effective, but which typically results in the underneath layer being washed out or dulled. With blending modes, you can create effects where the full intensity of the underlying layer is still visible. Blending mode settings can be found in the Symbology menu under **Layer Properties** in the **Layer Rendering** section (for vectors) and **Color rendering** section (for rasters).

Composing maps

With QGIS, you can create maps that can be printed or exported to image and graphic files. Prior to version 3, in order to create maps, you used the Print Composer; this has now changed its name to Layouts. To create a new Layout, click on **Project | New Print Layouts**. Give the new layout a name, click on **OK**, and the composer window will open. From this point on, you can access the new layout from **Project | Layouts**.

The composer presents you with a blank sheet of paper upon which you can craft your map. Along the left-hand side, there are a series of tools on the **Composer Items** toolbar. The lower portion of the toolbar contains buttons for adding map elements to your map. These include the map body, images, text, a legend, a scale bar, graphic shapes, arrows, attribute tables, and HTML frames. Map elements become graphics on the composition canvas. By selecting a map element, graphic handles will appear around the perimeter. These can be used to move and resize the element. The upper portion of the **Composer Items** toolbar contains tools for panning the map data, moving other graphic content, and zooming and panning on the map composition.

The majority of the map customization options can be found in the composer tabs. To specify the sheet size and orientation, use the **Composition** tab. Once map elements have been added to the map, they can be customized with the **Item properties** tab. The options available on the **Item properties** tab change according to the type of map element that is selected. The **Atlas generation** tab allows you to generate a map book. For example, a municipality could generate an atlas by using a map sheet GIS layer and specifying which attribute column contains the map sheet number for each polygon. The **Items** tab allows you to toggle individual map elements on and off.

The toolbars across the top contain tools for aligning graphics (the **Composer Items** actions toolbar), navigating around the map composition (the **Paper Navigation** toolbar), and tools for managing, saving, and exporting compositions. Maps can be exported as images, PDFs, and SVG graphic files. To export the map, click on the **Composer** menu and select one from among **Export as image...**, **Export as SVG...**, or **Export as PDF...** depending on your requirements. The following is a screenshot showing the **Layout** window:

Adding functionality with plugins

There are so many potential workflows, analysis settings, and datasets within the broad field of GIS that no out-of-the-box software could contain the tools for every scenario. Fortunately, QGIS has been developed with a plugin architecture. Plugins are add-ons to QGIS that provide additional functionality. Some are written by the core QGIS development team, and others are written by QGIS users.

You can also browse the QGIS Python Plugins Repository at `https://plugins.qgis.org/plugins/`.

You can explore the QGIS plugin ecosystem by navigating to **Plugins** | **Manage and Install Plugins**. This opens the **Plugins Manager** window (shown in the following screenshot), which will allow you to browse all plugins, those that are installed, those that are not installed, plugins that are newly available, and adjust the plugin manager settings. If there are installed plugins with available upgrades, there will also be an **Upgradable** option. The search bar can be used to enter search terms and find available plugins related to the topic. This is the first place to look if there's a tool or extra type of functionality that you need! To install a plugin, simply select it and click on the **Install Plugin** button. Installed plugins can be toggled on and off by checking the box next to each:

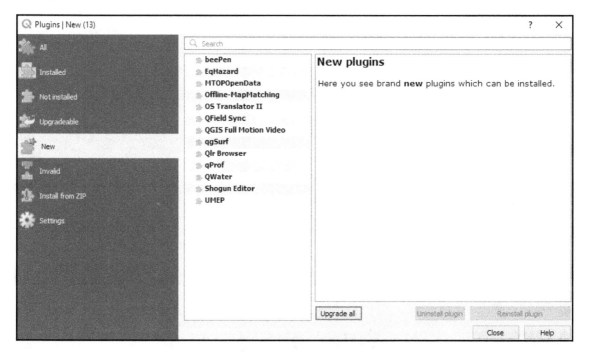

You will be notified by a link at the bottom of the QGIS Desktop application if there are updates available for your installed plugins. Clicking on the link will open the **Plugins Manager** window, where the **Upgrades** tab will allow you to install all or some of the available updates. Plugins themselves may show up as individual buttons, toolbars, or as items under the appropriate menu, such as **Plugins**, **Vector**, **Raster**, **Database**, **Web**, or **Processing**. The following information box describes a great plugin for adding base maps.

To add a basemap to QGIS, enable the **Quick Map Services** plugin. Once it's installed, it appears as an item on the **Web** menu. You can configure it to include more map services. From the menu bar, choose **Web** | **QuickMapServices** | **Settings**. Click on the **Contributed Services** tab, and click the **Get contributed pack** button. At this point, it will allow you to add base maps from OpenStreetMap, Google Maps, Bing Maps, Map Quest, OSM/Stamen, Apple Maps, and several more. This plugin requires an internet connection.

Custom QGIS variables

A new feature of QGIS since version 2.12 is custom variables. Variables can be set at several different scopes or levels:

- **Global**
- **Project**
- **Layer**

Global variables are available at the application level in any QGIS Desktop instance. These can be managed from **Settings** | **Options** | **Variables**. Examples of what you might set as global might be a company/organization name that you can then later use in expressions, Print Layouts, or when editing data to autopopulate fields.

Project variables are available throughout a given project. These can be set from the **Project Properties** window on the **Variables** tab. Examples may include project names, reference numbers, or client names.

Layer level variables are available from **Layer Properties** on the **Variables** tab, and are available only for that layer.

In each instance, variables can be added using the green plus button found in the bottom-right corner of the variables screens.

The other important aspect of these variables is that they cascade from most specific scope to least specific scope. In other words, @my_var set at the **Global** level will be overwritten by @my_var set at the **Project** level.

Variables are still a very new feature, but have a lot of potential for creating more efficient workflows. Once you begin using them, you will undoubtedly discover many other use cases.

Summary

This chapter provided a refresher on the basics of QGIS and a look at some of the new functionality of 3.x. We covered how to install the software on several platforms. We then covered how to load vector, raster, and database data layers. Next, you were shown how to work with coordinate reference systems and style data. We covered the basics of working with tables, including how to perform a table join. The chapter concluded with a refresher on composing maps, how to find, install, and manage plugins, and a primer on QGIS custom variables.

The next chapter will cover styling vectors and raster data. Now that you have had a refresher on the basics of QGIS, it is time to learn how to expand your work to include spatial databases. In Chapter 3, *Creating Spatial Databases*, you will learn how to create and manage spatial databases within QGIS.

Section 2: Getting Started

2

In this section, we will see how to style raster and vector data for display, create and edit spatial databases using QGIS, turn raw vector data into a more usable form, and then prepare raster data for further processing. This section will prepare you for advanced operations in QGIS.

In this section, we will cover following chapters:

- Chapter 2, *Styling Raster and Vector Data*
- Chapter 3, *Creating Spatial Databases*
- Chapter 4, *Preparing Vector Data for Processing*
- Chapter 5, *Preparing Raster Data for Processing*

Styling Raster and Vector Data

2

In this chapter, we will cover the advanced styling and labeling of raster and vector data in QGIS. It is assumed that you are familiar with basic styling in QGIS and are looking to improve your styling techniques.

The topics that we will cover in this chapter are as follows:

- Choosing and managing colors
- Managing color ramps
- Styling multiband rasters
- Raster color rendering
- Raster resampling
- Styling vectors
- Vector layer rendering
- Using diagrams to display thematic data
- Saving, loading, and setting default styles

Choosing and managing colors

As colors are used throughout the styling process, we will first review the ways in which you can select and manage color collections in QGIS. In the **Symbology** tab window (via the **Layer Properties**), do a single click on the color that is displayed next to the word color to access the **Select Color** window.

Alternatively, you can access a cut-down version of the same display by clicking on the down arrow at the end of the color selection, which is shown in the following screenshot:

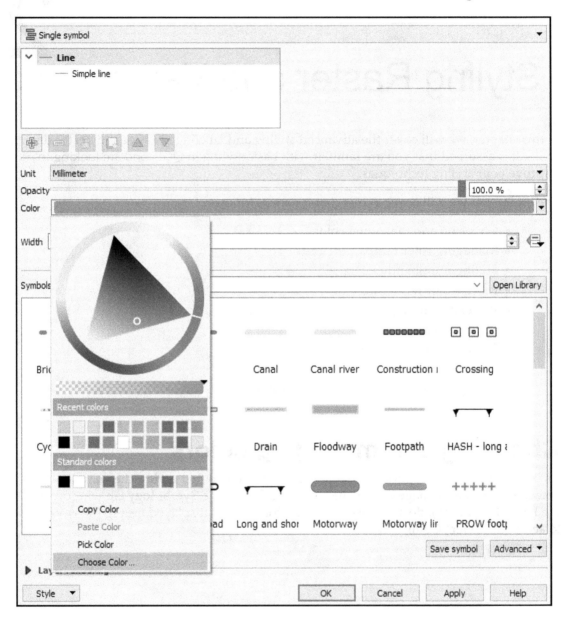

Accessing this slimmed-down version on the **Select Color** window gives you access to copy and paste colors. This is useful if you want to copy/paste colors between layers/features. If you click on **Choose Color...**, you can again access the **Select Color** window.

This will open the color picker tool, as shown in the following screenshot:

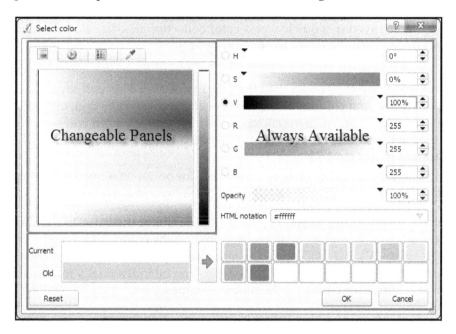

Let's take a tour of the color picker tool by starting with the components that are always available and conclude our tour by looking at the four changeable panels.

As of version 3.x of QGIS, you can also access the styling of layers via two other options:

- **Method one**: Right-click on a layer and go to the **Style** option (toward the bottom of the list). This will give you access to a color manager, as well as the ability to copy style, so that you can paste it against another layer.
- **Method two**: Go to **View| Panels** and choose **Layer Styling**. This will add a new panel on the left of your map canvas. With a layer selected in the **Layers** panel, click on the **Layer Styling** to have access to all of the same features as you could via the **Layer Properties | Symbology** method. While in the **Layer Styling** panel, you can switch between layers via the drop-down menu at the top of the window.

Knowing color picker components

Refer to the color picker tool screen to understand this section. The current and previous colors are displayed in the bottom-left corner of the color picker tool. In the **Select Color** window, the **Old** field depicts the color that is currently chosen, and the color mentioned in the **Current** field will replace the one in the **Old** field if the **OK** button is clicked. The current color can be saved in a quick-access color collection (the 16 colored squares in the bottom-right corner) using either of the following options:

- By clicking on the button with the blue arrow ( ): This will save the current color in the first column of the top row of the color collection, overwriting any existing color. Subsequent clicks on the blue arrow button will store the current color in the next column until all 16 boxes are full and will then loop back to the beginning.
- By dragging and dropping the current color on top of a quick-access color box: The old color can also be saved using the drag-and-drop method.

The color picker displays and allows for the manipulation of the value in the **Current** field in two color models: HSV and RGB. In the right half of the color picker, the hue (**H**), saturation (**S**), value (**V**), red (**R**), green (**G**), and blue (**B**) values for the currently selected color are displayed. Each of these color parameters can be individually modified by either using the slider controls or changing the numeric values.

> Red, green, and blue values must be specified between 0 and 255, where 0 represents no color and 255 represents full color. Hue is specified in degrees ranging from 0 to 359, where each degree represents a different location (and color) on the color wheel. Saturation and value are specified using percentages and range between 0%, representing no saturation or value, and 100%, representing full saturation or value.

Below the color parameters is the **Opacity** setting. The right halves of the **Current** and **Old** fields display the color with the applied opacity level. For example, in the following screenshot, the current color is shown with no opacity (100%) on the left and with the currently selected opacity of 50% on the right:

The **HTML notation** textbox displays the HTML color notation of the current color. The color notation can be changed to one of the four different formats by clicking on the down arrow (⏷) in the **HTML notation** textbox.

Lastly, the **Reset** button resets the current color to match the old color.

Changeable panels in the color picker

The color picker has four changeable panels: **Color ramp** ( ), **Color wheel** ( ), **Color swatches** ( ), and **Color sampler** ( ). Each of these panels provides convenient ways to select and manage colors. This section will provide the details of each of the four panels.

Color ramp

The **Color ramp** panel is an interactive selection tool that sets the currently selected HSV or RGB parameter values based on the location of a mouse click. To select the color model parameters that the color ramp will display, click on one of the radio selection buttons next to the **H**, **S**, **V**, **R**, **G**, or **B** values in the right half of the color picker. The selected parameter can be individually modified using the thin vertical slider control on the right-hand side of the **Color ramp** panel. The other two color model parameters can be set simultaneously by clicking on the large color display on the left-hand side of the color ramp panel. In the following screenshot, the **V** (Value) value in the HSV color model is selected and is represented in the thin vertical slider control; the **H** (Hue) and **S** (Saturation) values are combined and are represented in the large color display:

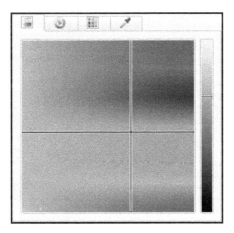

Color wheel

The **Color wheel** panel is an interactive selection tool that sets the color value based on mouse clicks. The ring contains the hue, while the triangle contains the saturation and value. To set the hue, click on the ring. To set the saturation and value (while not changing the hue), click on the triangle:

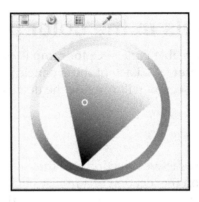

Color swatches

The **Color swatches** panel provides an interface to manage color palettes and select colors from the palettes:

To switch between color palettes, select the desired color palette from the drop-down box at the top. The following three default color palettes are listed in the drop-down box:

- **Recent colors**: This contains the colors that have been selected the most recently in the color picker. This palette cannot be modified.
- **Standard colors**: This contains the colors that are always available as quick selections in QGIS.
- **Project colors**: This contains the colors that are stored within the QGIS project file.

The three default color palettes can be quickly accessed by clicking on the down arrow next to a color display, as shown in the following screenshot:

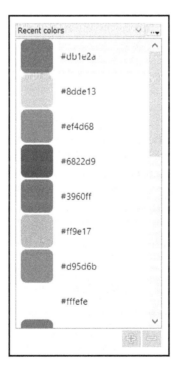

The options drop-down button to the right of the Recent colors provides several handy functions to manage palettes. Let's review each one:

- **Copy Colors**: This copies the selected color(s) in the current palette to the clipboard.
- **Paste Colors**: This pastes the color(s) stored in the clipboard to the current palette.

- **Import Colors**: This imports colors from a GIMP Palette (GPL) file and places them into the current palette.

 A GPL file is a text file that stores color palette information. The GPL file format is from the free and open source GIMP image editor project. More information about GIMP can be found here: http://www.gimp.org.

- **Export Colors**: This exports the current palette to a GPL palette file.
- **New Palette**: This creates a new, empty palette that you can name. The palette will persist in the color picker until it is removed.
- **Import Palette**: This imports a GPL palette file into the list of palettes. The palette will persist in the color picker until it is removed.
- **Remove Palette**: This removes the current palette from the list. Note that the three default palettes cannot be removed.

Color sampler

The **Color sampler** sets the current color to a color sample that's been collected from the screen using the mouse pointer. The color sample is based on the average of all colors under the mouse pointer within the specified **Sample average radius** value. To collect a sample color, click on the **Sample color** button, then move the mouse cursor to a location where you want to sample a color, and either press the spacebar or click to collect the sample. As you move the mouse cursor around, a preview of the sample color will appear under the **Sample color** button. The following screenshot shows the color sampler with **Sample average radius** of 5 px (pixels) and a preview of the green color currently under the mouse cursor:

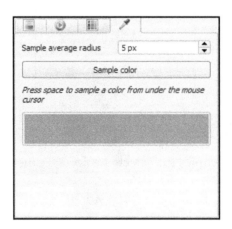

Now that the color picker dialog has been toured and you know how to select colors and manage them in palettes, we will look at how to create and manage color ramps.

Color picker components in the Layers panel

For vector layers, the color wheel and recent colors can be accessed through a context menu in the **Layers** panel. Depending on the type of style applied to the vector layer (vector layer styles are discussed in detail later in this chapter), the color wheel will be accessible in different ways.

If the vector layer has a single symbol style, the color wheel can be accessed by right-clicking on the layer name in the **Layers** panel, then choosing **Styles** to expose the color wheel and recent colors (shown in the following screenshot):

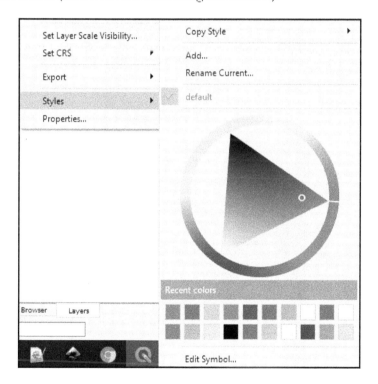

Further to this, as of version 3.x QGIS now has the Layer Styling Panel that allows you quick access to a view of the **Symbology** tab. To add this feature to the Panel, go to **View** | **Panels** | **Layer Styling Panel**. In the Layer panel simply click on a layer and then click on the Layer Styling Panel. In addition to this, the styling changes that you make in the Layer Styling Panel are rendered Live, is as much as they update the data on the Map Canvas in realtime without the need to Apply changes.

To change which layer to render, you can either tab back into the **Layers** panel and choose the layer to change or, at the top of the **Layer Styling Panel,** you can access GIS layers via a drop-down option:

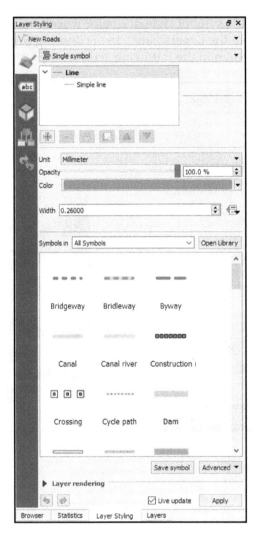

If the vector layer uses **Categorized**, **Graduated**, or **Rule-based** styles, as we will see soon in the *Styling Vectors* section, then the color wheel and recent colors can be accessed by right-clicking on an individual symbol from the expanded layer in the **Layers** panel:

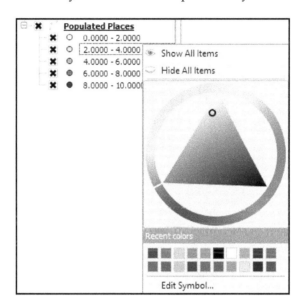

With the color wheel and recent colors displayed in the context menu, the color changes you make in the context menu will be changed on the map canvas in real time. To see all the symbol options for the selected symbol or class, click on **Edit Symbol**.

Managing color ramps

Color ramps are used in multiple applications when styling data. A color ramp is a series of continuous or discrete colors that can be applied to raster or vector data values. QGIS contains a number of color ramps that are ready to use and allows you to add new color ramps. In this section, we will first demonstrate how to manage the QGIS color ramp collection and then how to add new color ramps.

Managing the QGIS color ramp collection

Color ramps can be managed and created using the **Style Manager** window. The **Style Manager** window provides an interface to manage the **Marker**, **Line**, **Fill**, and **Color ramp**, which are available in the **Style** tab of the **Layer Property** window.

To open the Style Manager, navigate to **Settings | Style Manager** and then click on the **Color ramp** tab. The **Style Manager** window is shown in the following screenshot:

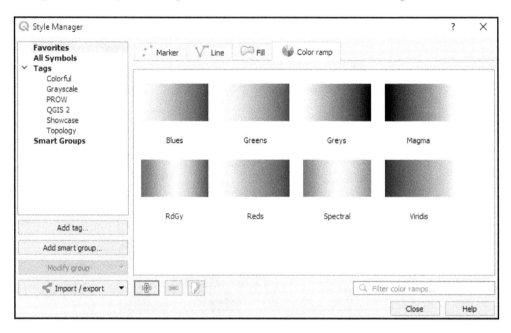

The color ramps that are displayed in the **Color ramp** tab are available for quick access from drop-down selection boxes when you style the data.

For example, the following screenshot shows the quick-access color ramps in a drop-down box when we specify a color ramp to apply a **pseudocolor** to a **singleband** raster:

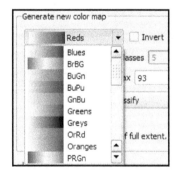

Six operations are available to manage color ramps in the **Style Manager**: **rename**, **remove**, **export**, **import**, **add**, and **edit**. Each of these operations will be explained now.

Renaming a color ramp

To rename a color ramp, click once on the color ramp to select it, pause, and then click on it a second time (this is a slow double-click) to make the name editable. Type in the new color ramp name; then, press *Enter* to save it.

Removing a color ramp

To remove a color ramp, select the color ramp and then click on the **Remove item** button ( ). The color ramp will no longer be available.

Exporting a color ramp

To export a color ramp, click on the **Share** button ( ) in the lower-right corner. Then, click on **Export** to open the **Export symbol(s)** window as shown in the following screenshot:

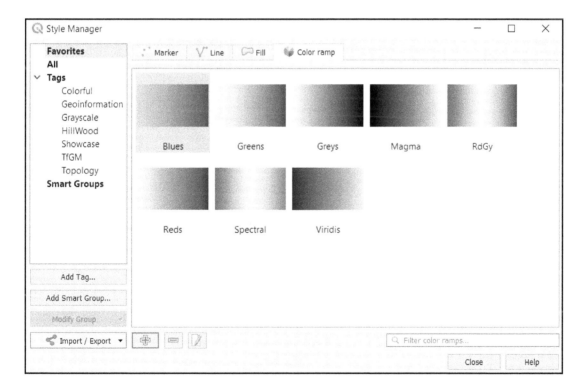

Select the symbols that you wish to export and then click on **Export** to export the selected symbols to an XML file. The exported symbols can later be imported to the **Style Manager** using the **Import** function.

Importing a color ramp

To import color ramps from an XML file, click on the **Share** button (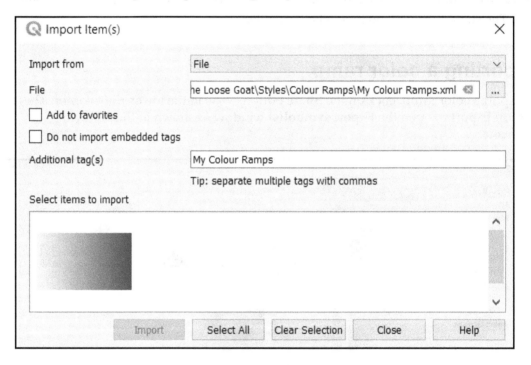) in the lower-right corner; then, click on **Import** to open the **Import symbol(s)** window:

Exported color ramp styles can be imported from an XML file or a URL that is pointing to an XML file.

To import from a file, select **file specified below** for the **Import from** parameter; then, click on **Browse** to select the XML file. Once the **Location** value is specified, the color ramps will be displayed. Select the color ramps that you wish to import, select the group into which you wish to import the color ramps, and then click on **Import**. The imported color ramps (and other symbol types, if selected) will be added to the color ramp list in the **Style Manager**.

Adding a color ramp

Using the **Add item** button (⊕▾), three types of color ramps—*Gradient, Random,* and *ColorBrewer*—can be added to the color ramp list in the **Style Manager**. These color ramp types can be created from scratch, or they can be selected from a large collection of existing color ramps from the cpt-city archive of color gradients.

Let's add one color ramp of each type and then add a cpt-city color ramp.

Adding a gradient color ramp

To add a gradient color ramp, click on the **Add item** button (⊕▾) and then choose **Gradient**. This will open the **Gradient Color Ramp** window. A gradient color ramp uses two colors, which are specified as **Color 1** and **Color 2**, to set the start and end colors. QGIS applies an algorithm to create a gradation between the two colors.

The following screenshot shows the **Gradient Color Ramp** window:

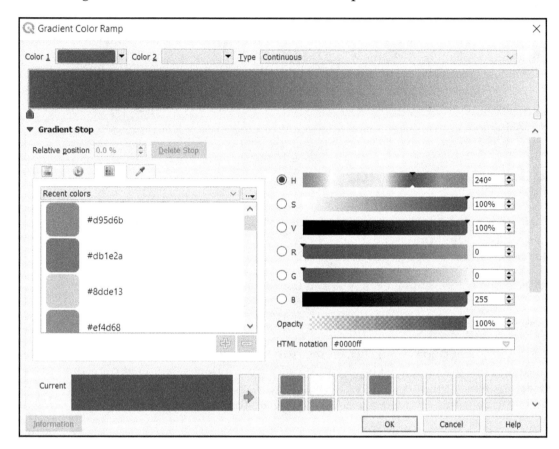

More colors can be added by double-clicking inside the color ramp to add a new **Gradient stop**.

Each **Gradient stop** can be moved along the **Color Ramp**, along with its actual color and transparency:

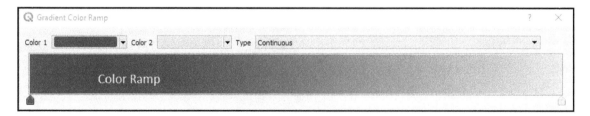

When all of the gradient parameters have been set, click on **OK** to save the gradient. QGIS will prompt you to name the gradient and to give it any **Tag**(s) that you wish. Once it is named, the gradient will appear in the Style Manager's list of color ramps after the default color ramps.

Adding a random color ramp

A **Random color ramp** generates a number of randomly generated colors that fall within specified **Hue**, **Saturation**, and **Value** ranges.

To add a **Random color ramp**, click on the **Add item** button ( ) and then choose **Random**. This opens the **Random color ramp** window. The **Classes** parameter determines how many colors to generate. Colors are randomly generated each time any of the parameters are changed. As an example, in the following screenshot, five random colors are generated with different hues (between 100 and 320) but with the same saturation and value:

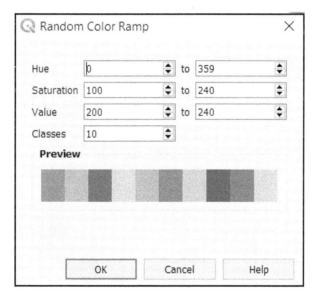

When all of the parameters have been set, click on **OK** to save the random color ramp. QGIS will prompt you to name the color ramp. Once it is named, the **Random color ramp** will appear in the Style Manager's list of color ramps.

Adding a ColorBrewer color ramp

To add a **ColorBrewer ramp**, click on the **Add item** button ( ) and then choose **ColorBrewer**. This opens the **ColorBrewer ramp** window (shown in the following screenshot). A **ColorBrewer ramp** generates three to eleven colors using one of the available schemes. The **Colors** parameter determines how many colors to generate, and the **Scheme name** parameter sets the color scheme that will be used. As an example, in the following screenshot, five colors are generated using the **RdGy** (Red to Grey) color scheme:

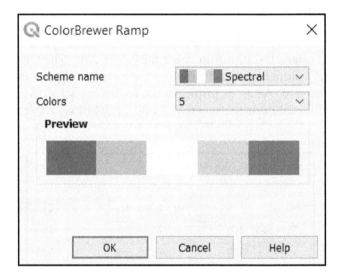

When all of the parameters have been set, click on **OK** to save the **ColorBrewer ramp**. QGIS will prompt you to name the color ramp. Once it is named, the **ColorBrewer color ramp** will appear in the Style Manager's list of color ramps.

> The ColorBrewer color ramps are based on the work of Cynthia Brewer. For more information and an interactive color selector, visit the ColorBrewer website at `http://colorbrewer2.org/`.

Adding a cpt-city color ramp

If you do not want to add a color ramp from scratch, a large collection of existing color ramps from the cpt-city archive of color gradients is available for use in QGIS. To add a **cpt-city color ramp**, click on the **Add item** button (⬤▾) and then choose **Catalog: cpt-city**. This opens the **cpt-city color ramp** window:

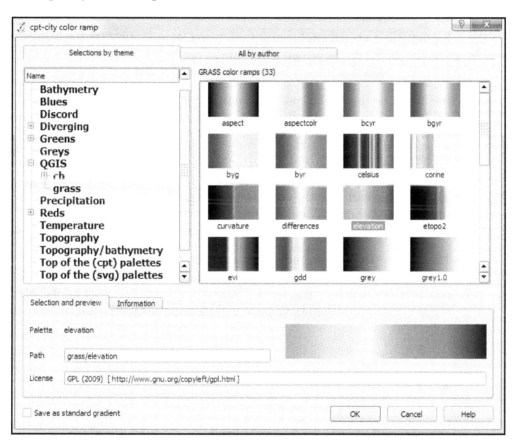

Color ramps can be selected by theme or by author by choosing the appropriate tab at the top of the window. In either case, the color ramps are presented in an expandable tree on the left, with a list of color ramps in each tree element on the right. When a color ramp is selected, the **Selection and preview** and **Information** tabs are populated.

There are two ways to add a **cpt-city color ramp** to the list in the **Style Manager**: as a cpt-city or as a standard **Gradient color ramp**.

To save the color ramp as a **cpt-city color ramp**, click on **OK** with a color ramp selected. This will keep the link between the added color ramp and the **cpt-city color ramp** list. The color ramp cannot be modified if it is added as a **cpt-city color ramp**.

Be sure to review the license information for the **cpt-city color ramps** in the **Information** tab's **License** field. Many different licenses are used and some require attribution before they can be used.

To save the color ramp as a standard gradient color ramp, check **Save as standard gradient** and then click on **OK**. This will save the color ramp as a gradient color ramp, and the color ramp will not link back to the cpt-city color ramp collection. The color ramp can be modified later as it has been converted in to a standard gradient color ramp.

The cpt-city archive of gradients is available at `http://soliton.vm.bytemark.co.uk/pub/cpt-city/`. The archive contains thousands of gradients. The gradients that are most applicable to styling geographic data have been included in the QGIS cpt-city collection.

Editing a color ramp

To edit a color ramp, select the color ramp and then click on the **Edit item** ( ) button. This will open one of the four types of windows, depending on which type of color ramp was selected: Gradient, Random, ColorBrewer, or cpt-city. Using the opened window, the properties of the color ramp can be modified.

Now that color ramps have been discussed, we will put the color ramps to work by styling raster data with color ramps. Later, we will use color ramps to style vector data.

Styling singleband rasters

In this section, the three different band render types that are appropriate for singleband rasters will be covered. Singleband rasters can be styled using three different band render types: paletted, singleband gray, and singleband pseudocolor.

Note that even though raster color rendering and resampling are part of raster style properties, they will be discussed separately in later sections as they are common to all singleband and multiband raster renderers.

> The raster band render type should be chosen to best match the type of data. For instance, a palette renderer is best used on rasters that represent discrete data, such as land use classes. The singleband gray would be a good choice for a hillshade, while a singleband pseudocolor would work well on a raster containing global temperature data.

Paletted raster band rendering

The paletted raster band renderer applies a single color to a single raster value. QGIS supports the loading of rasters with paletted colors stored within and changing the color assigned to the raster value. QGIS does not currently support the creation of color palettes for singleband rendering. However, existing QGIS layer style files (.qml) that contain palettes can be applied by clicking on the **Load Style** button in the layer properties.

As an example of a raster with a color palette stored within it, add SR_50M_alaska_nad.tif from the raster folder in the sample data, to the QGIS canvas and open the **Symbology** tab under **Layer Properties**. The following figure shows the **Paletted / Unique Value** renderer being applied to **Band 1 (Gray)** with a Color Ramp of Reds:

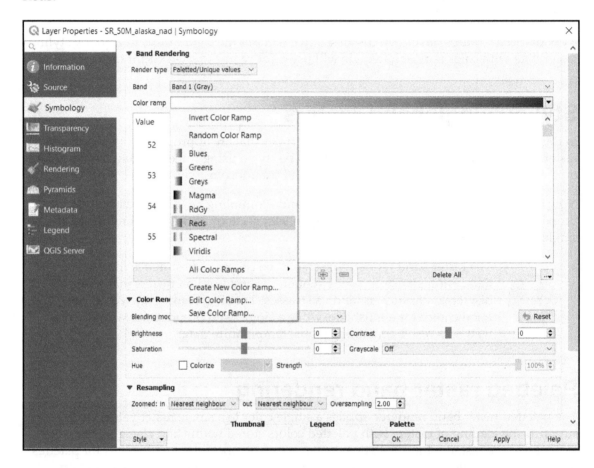

This is the output:

To change a color, double-click on a color in the **Color ramp** column to open the color picker.

Singleband gray raster band rendering

The singleband gray band renderer stretches a gradient between black and white to a single raster band. Additionally, contrast enhancements are available to adjust the way the gradient is stretched across the raster band's values. Let's apply a singleband gray renderer and a contrast enhancement to the same sample raster, Gray_50M_SR_W.tif, which represents shaded relief, hypsography, and flat water for Earth.

Add Gray_50M_SR_W.tif to the QGIS canvas and open its **Symbology** tab from **Layer Properties**. As this is a singleband raster, QGIS defaults the **Render type** value to **Singleband gray** with the following parameters:

- **Gray band**: **Band 1 (Gray)**. This is the raster band that is being styled. If a multiband raster is being used, then the combobox will be populated with all raster bands.
- **Color gradient**: **Black to white**. This is the gradient to apply to the selected gray band. The choices are **Black to white** and **White to black**.
- **Min**: 56. The minimum cell value found in the gray band.

- **Max**: 250. The maximum cell value found in the gray band.
- **Contrast enhancement**: **Stretch to MinMax**. The method that's used to stretch the color gradient to the gray band with respect to the **Min** and **Max** values.

These are shown in the following screenshot:

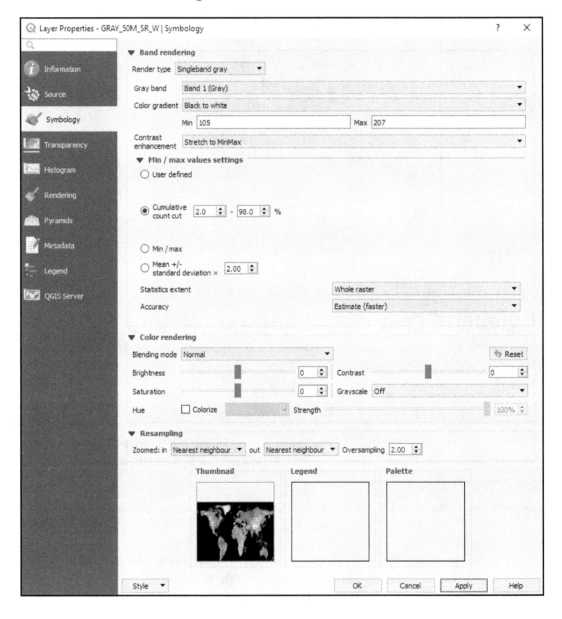

Min / max value settings: This selects cell values to include in the **Min** and **Max** value determination. Rasters may have cell values that are outliers, which may affect the rendering of the image. For instance, if only a few cells have an abnormally high value, then the gradient will stretch all of the way to these high values, which will cause the raster to look overly gray and bland. To combat this grayness, some cell values can be excluded so that the gradient is not skewed by these outliers. Three methods are available to select cell values, and it is recommended that you experiment with these values to achieve the most desirable selection of cell values:

- **Cumulative count cut**: This includes all values between the two parameters. In the preceding screenshot, all values between 2% and 98% of the cell data range were included. In general, this will remove the few very high and very low values that may skew the gradient.
- **Min / max**: This includes all values.
- **Mean +/- standard deviation**: This includes all values within the specified number of standard deviations about the mean of all values.
- **Statistics Extent**: The extent of the raster to sample for cell values. Either the **Whole raster**, extent of the raster, or the **Current or Updated** canvas extent can be used.
- **Accuracy**: This determines the accuracy of the min/max calculation. The calculation can either be an **Estimate (faster)** or an **Actual (slower)** option. In general, **Actual (slower)** is the preferred option; however, for very large rasters, **Estimate (faster)** may be preferred to save time.

With the **Load min/max values** set, click on the Apply button to calculate the **Min** and **Max** values. With the **Min** and **Max** values set, we can turn our attention to the **Contrast enhancement** parameter. The **Contrast enhancement** parameter sets how to stretch the color gradient across the cell values of the gray band. The following four methods are available for **Contrast enhancement**:

- **No enhancement**: No enhancement is applied. The color gradient is stretched across all values in the entire gray band. While this may be desired sometimes, it may tend to make the raster look overly gray.
- **Stretch to MinMax**: This method stretches the color gradient across the gray band between the **Min** and **Max** values. It generally produces a higher contrast, a darker rendering than **No enhancement**. All cell values below the **Min** value are assigned the lowest gradient color and all cell values above the **Max** value are assigned the highest gradient color.

- **Stretch and clip to MinMax**: This method stretches the color gradient across the gray band between the **Min** and **Max** values. It produces the same rendering as the **Stretch to MinMax** method, except that all cell values below the **Min** value and all values above the **Max** value are assigned no color (and they are transparent).
- **Clip to MinMax**: This method stretches the color gradient across all values in the gray band, which is the same result as **No enhancement**, except that all cell values below the **Min** value and all values above the **Max** value are assigned no color (and they are transparent).

The following screenshot shows the effects of the four different **Contrast enhancement** methods on the Gray_50M_SR_W.tif sample file when the **Color gradient** field is set to **Black to white**, **Min** is set to 107, and **Max** is set to 207. A **Min** value of 107 is selected to exclude the cell value of 106 that is associated with the oceans:

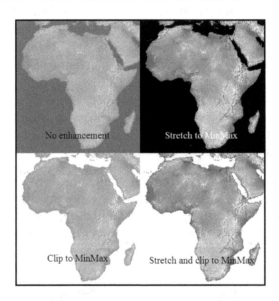

Singleband pseudocolor raster band rendering

The singleband pseudocolor band renderer stretches a color ramp to a single raster band. Additionally, three **Color interpolation** methods are available to adjust the way the color ramp is stretched across the raster band's values with respect to the min and max cell values (for a discussion on determining min and max values, see the preceding section).

Let's apply a singleband pseudocolor renderer to the GRAY_50M_SR_W.tif sample data raster file that represents shaded relief, hypsography, and flat water for Earth. Add GRAY_50M_SR_W.tif to the QGIS canvas and open its **Symbology** tab from **Layer Properties**. For the **Render type** field, choose **Singleband pseudocolor**.

The singleband pseudocolor render type has many interworking parameters that are best explained as a whole through the lens of a workflow, instead of explaining them as separate parts.

The example that's shown in the following screenshot will be the basis for explaining the parameters:

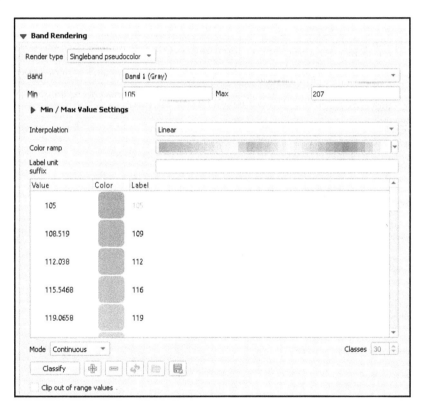

Using the preceding screenshot for reference, complete the following steps:

1. Go to **Settings | Style Manager** and click on the **Color ramp.**

2. Click Add and choose **catalog: cpt-city**.

3. Navigate to the **Topography/bathymetry** section on the left and choose **wiki-2.0**.

4. Give it a name, for example, `Atlas_Style`, and click **Save**, then close the window.
5. Go to the **Layer Properties** and **Symbology** tab.
6. Click in the drop-down menu for the **Color ramp**, go to **All Color Ramps**, and choose `Atlas_Style`. If you have not done so already, change the **Band Min** value to `105` and set the **Max** value to `207`.
7. Click on the **Classify** button to apply the color ramp to the values. The classification list on the left will populate with values, colors, and labels.
8. Click **Apply**.

Just for your reference, refer the following screenshot:

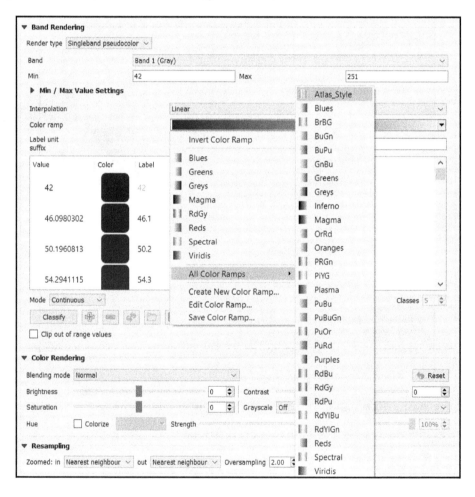

Singleband pseudocolor interpolations

The following interpolations fall under the singleband pseudocolor option. Let's have a quick look at them:

- **Linear**: By default, QGIS displays the raster under the singleband pseudocolor option with an **Interpolation** set to **Linear**. It assigns a unique color to each unique raster value. Values between values that are listed in the **Value** column are assigned a unique color that is calculated linearly and is based on its location between the surrounding listed values. In other words, if there are, say, *164* unique values in the raster and *15* colors listed in the classification list, the raster will be rendered with the *164* unique colors that appear as a nice, linear progression through the *15* listed colors. This method is best for raster data that represents continuous information (for example, elevation or temperature data) where you want a smooth progression of color that stretches across the raster values. If you change this value to **Discrete,** then click on **Classify** and then **Apply**; the rendering of the raster will change.

- **Discrete**: Assigns only, and exactly, the colors that were chosen in the classification list. Values between values that are listed in the **Value** column are assigned the color that are assigned to the next highest listed value. In other words, if there are, say, *164* unique values in the raster and *15* colors listed in the classification list, the raster will be rendered with exactly the *15* listed colors. This method is best for cases where you want to reduce the number of colors that will be used to render the raster. Furthermore, you can change the value to **Exact**, then click on **Classify**, and then **Apply**.

- **Exact**: This assigns a unique color to only the values listed in the **Value** column of the classification list. In other words, if there are, say, *164* unique values in the raster and *15* colors (and *15* associated values) listed in the classification list, only the *15* raster values that are listed will be rendered with their associated colors. No other values will be assigned a color. This method is best for raster data that represents discrete data classes where you do not want non-listed values to be assigned any color.

The following screenshot shows the effects of the three **Color interpolation** methods on our sample data, as configured so far:

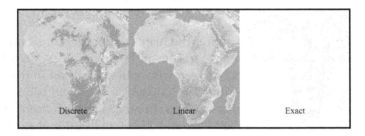

Styling multiband rasters

The multiband color band renderer stretches three gradients (red, green, and blue) to three separate raster bands. The basic idea is that the computer will display natively used combinations of red, green, and blue lights to create the desired image. By matching individual raster bands to the red, green, and blue lights that are used by the display, the three bands' colors will mix so that they are perceived as other colors, thereby creating a red, green, and blue image composite that is suitable for display.

Contrast enhancements are available to adjust the way the gradients are stretched across the raster bands' values. Contrast enhancements have already been covered in the *Singleband gray raster band rendering* section, so refer to this section for an in-depth coverage of the topic.

Raster color rendering

Raster color rendering modifies the properties of the raster to change the way it displays and interacts with the layer below it in the **Layers** panel. Color rendering is a part of the raster style properties for all band renderer types and works in the same way, regardless of the selected band renderer. In this section, we will discuss the parameters that are available for change in the **Color rendering** section of the raster style properties.

When a raster is first loaded, the **Color rendering** parameters are set to their default values, as shown in the following screenshot. At any time, the default values can be reloaded by clicking on the **Reset** button:

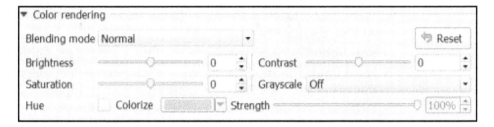

There are seven parameters that can be set in the **Color rendering** section, which are as follows:

- **Blending mode**: This applies a blending method to the raster that mixes with layers below it in the **Layers** panel. A number of blending modes are available to choose from and these are commonly found in graphics editing programs. There are 13 blending modes, as follows:
    - **Normal**: This is the default blending node. If the raster has any transparent cells, the colors from the layer below the raster will show through; otherwise, no colors will be mixed.
    - **Lighten**: In this mode, for each raster cell in the or the raster below, the maximum value for each color component that is found in either raster is used.
    - **Screen**: In this mode, lighter cells from the raster below are displayed transparently over the raster while darker pixels are not.
    - **Dodge**: This mode increases the brightness and saturation of the raster cells below based on the brightness of the raster's cells.
    - **Addition**: This mode adds the color components of each cell of this raster and the raster below together. If the color component value exceeds the maximum allowed value, then the maximum value is used.
    - **Darken**: In this mode, for each raster cell in the raster, or the raster below, the minimum value of each color component found in either raster is used.
    - **Multiply**: This mode multiplies the color components of each cell of the raster and the raster below together. This will darken the raster.

- **Burn**: In this mode, the raster below is darkened using the darker colors from this raster. Burn works well when you want to apply the colors of this raster subtly to the raster below it.
- **Overlay**: This mode combines the multiply and screen methods. When this is used for the raster below, lighter areas become lighter and darker areas become darker.
- **Soft light**: This mode combines the burn and dodge methods.
- **Hard light**: This mode is the same as the overlay method; however, this raster and the raster below are swapped for inputs.
- **Difference**: This mode subtracts this raster's cell values from the cell values of the raster below. If a negative value is obtained, then the cell value from the raster below is subtracted from this raster's cell value.
- **Subtract**: This mode subtracts this raster's cell values from the cell values of the raster below. If a negative value is obtained, a black color is displayed.

- **Brightness**: This changes the brightness of the raster. Brightness affects how bright or dark the raster appears. Brightness affects all cells in the raster in the same way.
- **Contrast**: This changes the contrast value of the raster. **Contrast** separates the lightest and darkest areas of the raster. An increase in contrast increases the separation and makes darker areas darker and brighter areas brighter. For example, a large negative contrast of -75 would produce a mostly gray or monotone image, since the bright and dark colors are not separated very much at all.
- **Saturation**: This changes the saturation value of the raster. Saturation increases the separation between colors. An increase in saturation makes the colors look more vibrant and distinct, while a decrease in saturation makes the colors look duller and more neutral.
- **Grayscale**: This renders the raster using a grayscale color ramp. The following three rendering methods are available:
    - **By lightness**: In this method, an average of the lightness value of multiple raster band values will be applied to the gray color ramp with the saturation set as 0. If the raster only has one band, then each cell's lightness value will be used. The lightness value is calculated using the formula: *0.5 * (max(R,G,B) + min(R,G,B))*.

- **By luminosity**: In this method, a weighted average of multiple raster band values will be applied to the gray color ramp. Luminosity approximates how you perceive brightness from colors. The weighted average is calculated using the formula: *0.21 * red + 0.72 * green + 0.07 * blue*.
- **By average**: In this method, the average of the raster band values for each cell will be applied to the color ramp. If the raster only has one band, this selection will have no effect. For example, if the raster had three bands with cell values of 25, 50, and 75, then 50 would be applied as the cell value for the gray color ramp. The average is calculated using the formula: *(R+G+B)/3*.

- **Hue**: This parameter adds a hue to each cell of the raster. To apply a hue, check the **Colorize** box and then select a color using the color picker.
- **Strength:** This parameter linearly scales the application of the selected **Colorize** color to the existing raster colors.

Raster resampling

Raster resampling prepares the raster for display when not every raster cell can be mapped to its own pixel on the display. If each raster cell is mapped to its own display pixel, the raster renders at full resolution (also known as 1:1). However, since screen sizes are limited and we may wish to enlarge or reduce the size of the raster as we work at different map scales, the raster cells must be mapped to more than one pixel or a number of raster cells must be combined, or dropped, to map to a single pixel. As some raster cells cannot be shown at different resolutions, QGIS must determine how to render the raster and still maintain the character of the full-resolution raster. This section will discuss the parameters that are available for determining how the raster will be resampled for display.

The **Resampling** section of the raster **Style** tab has three parameters: **Zoomed: in**, **Zoomed: out**, and **Oversampling**.

The **Resampling** section with its default parameters is shown in the following screenshot:

Let's have a look at these parameters:

- **Zoomed: in** parameter sets the resampling method when zoomed in on the raster. Three resampling methods are available for selection: **Nearest neighbour**, **Bilinear**, and **Cubic**.
- **Zoomed: out** parameter sets the resampling method when zoomed out from the raster. Two resampling methods are available for selection: **Nearest neighbour** and **Average**.
- **Oversampling** parameter determines how many subpixels will be used to compute the value when zoomed out.

The four resampling methods that can be selected for use are as follows:

- **Nearest neighbour**: In this method, each raster cell is assigned the value of the nearest cell (measure between cell centers). This is a great method to choose when the raster represents discrete, categorical data as no new values are created.
- **Bilinear**: In this method, each raster cell is assigned an average value based on the four closest cells with original values. This method will smooth the data and may flatten peaks and fill valleys.
- **Average**: In this method, each raster cell is assigned an average value based on surrounding cells with original values. This method will smooth the data and may flatten peaks and fill valleys.
- **Cubic**: In this method, each raster cell is assigned an interpolated value based on the surrounding cells with original values. Unlike the bilinear method, this method will not smooth the peaks or valleys as much, and it tends to maintain local averages and variability. This is the most computationally intensive method.

Styling vectors

In this section, the eight different vector styling types will be covered. The eight types are single-symbol, categorized, graduated, rule-based, point-displacement, point-cluster, heat map, inverted polygon, and 2.5 D.

Single-symbol vector styling

The single-symbol vector style applies the same symbol to every record in the vector dataset. This vector style is best when you want a uniform look for a map layer, such as when you style lake polygons or airport points.

The following screenshot shows the **Single Symbol** style type with default parameters for point vector data. Its properties will be very similar to line and polygon vector data:

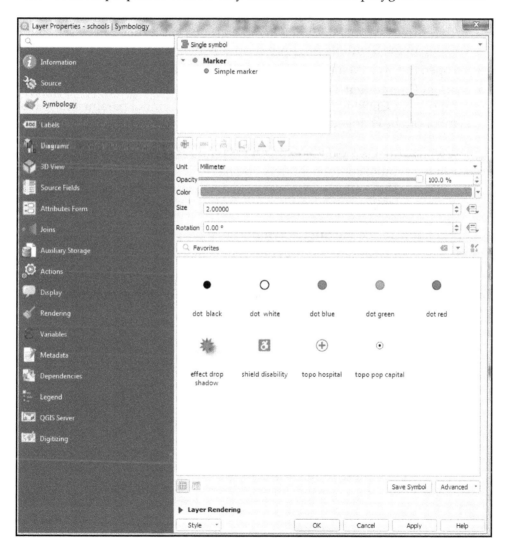

Let's take a quick tour of the four parts of the properties window:

- Symbol **Preview**, in the upper-left corner, shows a preview of a symbol with the current parameters.
- Symbol **Parameters**, in the upper-right corner, has the parameters for the symbol that's selected in the symbol component tree (these will change slightly depending on the geometry type of the vector data).
- **Library Symbols**, in the bottom-right corner, lists a group of symbols from the library (which is also known as the **Style Manager**). Clicking on a symbol sets it as the current symbol design. If symbol groups exist, they can be selected for viewing in the **Symbols in group** drop-down menu.
- To open the Style Manager, click on **Open Library**.
- **Symbol Component Tree**, in the bottom-left corner, lists the layers of symbol components. Clicking on each layer changes the symbol parameters so that the symbol can be changed.

 When you have a sub-component (**Simple marker**, for example) selected in the symbol component tree, you have the option to enable **Draw effects**. These effects are covered in detail later in this chapter in the *Draw effects* section.

As an example of how to use the **Single Symbol** style properties to create a circle around a gas pump ⬤, a second layer with the **SVG marker** symbol layer type can be added by clicking on the **Add symbol layer** button, and then moved on top of the circle by clicking on the **Move up** button (⬛). The following screenshot shows the parameters that are used to create the symbol:

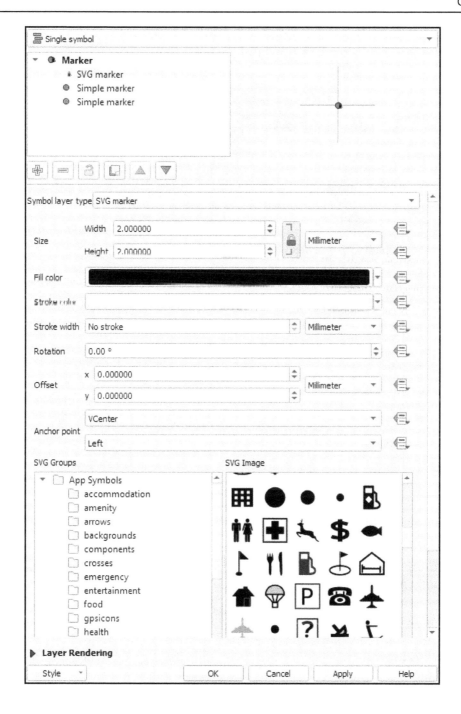

To save your custom symbol to the **Style Manager**, click on the **Save** button to name and save the style. The saved style will appear in the **Style Manager** and the list of library symbols.

Categorized vector styling

The categorized vector style applies one symbol per category of the attribute value(s). This vector style is best when you want a different symbol that is based on attribute values, such as when styling country polygons or classes of roads lines. The categorized vector style works best with nominal or ordinal attribute data.

The following screenshot shows the **Categorized** style type with parameters for point vector data of schools. Its properties will be very similar to those for line and polygon vector data:

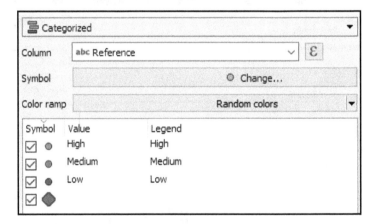

Styling vector data with the **Categorized** style type is a three-step process, which is as follows:

1. **Column:** Select an appropriate value for the field to use the attributes for categorization.
2. **Classify:** Create the classes to list by clicking on the button to add a class for each unique attribute that is found; otherwise
3. **Add:** This button adds an empty class. You can then double-click in the **Value** column to set the attribute value that's going to be used to create the class.

Classes can be removed with the **Delete** or **Delete all** buttons. They can be reordered by clicking and dragging them up and down the list. Classes can also be modified by double-clicking in the **Value** and **Legend** columns. Set the symbol for all classes by clicking on the **Symbol** button to open the **Symbol selector** window. Individual class symbols can be changed by double-clicking on the **Symbol** column of the class list. Choose the color ramp to apply to the classes. Individual class colors can be changed by double-clicking on the **Symbol** column of the class list.

Under the **Layer rendering** section, you can change the **Opacity** for all symbols, as well as change the **Blending mode**, which we discussed in the previous chapter. **Draw effects** enabled users to add shadows, glows, and various other effects.

Additionally, advanced settings are available by clicking on the **Advanced** button.

Graduated vector styling

The graduated vector style applies one symbol per range of numeric attribute values. This vector style is best when you want a different symbol that is based on a range of numeric attribute values, such as when styling gross domestic product polygons or city population points. The graduated vector style works best with ordinal, interval, and ratio numeric attribute data.

The following screenshot shows the **Graduated** style type with parameters for a point dataset that contains surveying levels:

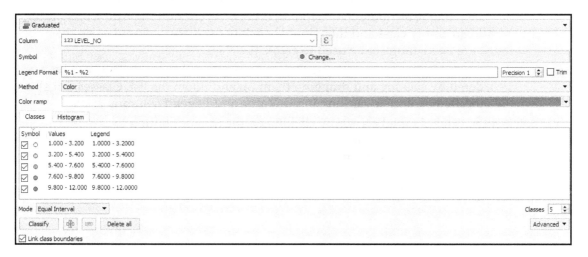

Styling vector data with the **Graduated** style type is a five-step process, which is as follows:

1. Select an appropriate value for the **Column** field to use the attributes for classification. The following five modes are available for use:
   - **Equal Interval**: In this mode, the width of each class is set to be the same. For example, if input values ranged between *1* and *100* and four classes were desired, then the class ranges would be *1-25, 26-50, 51-75,* and *76-100,* so that there were *25* values in each class.
   - **Quantile (Equal Count)**: In this mode, the number of records in each class is distributed as equally as possible, with lower classes being overloaded with the remaining records if a perfectly equal distribution is not possible. For example, if there are fourteen records and three classes, then the lowest two classes would contain five records each and the highest class would contain four records.
   - **Natural Breaks (Jenks)**: The Jenks Breaks method maximizes homogeneity within classes and creates class breaks that are based on natural data trends.
   - **Standard Deviation**: In this mode, classes represent standard deviations above and below the mean record values. Based on how many classes are selected, the number of standard deviations in each class will change.
   - **Pretty Breaks**: This creates class boundaries that are round numbers to make it easier for humans to delineate classes.

2. Create the classes to list by either clicking on the **Classify** button to add a class for each unique attribute that is found; otherwise, click on the **Add Class** button to add an empty class and then double-click in the **Value** column to set the attribute value range to be used to create the class.

3. Classes can be removed with the **Delete** or **Delete All** buttons. They can be reordered by clicking and dragging them up and down the list. Classes can also be modified by double-clicking in the **Value** and **Legend** columns.

4. Set the symbol for all classes by clicking on the **Symbol** button to open the **Symbol selector** window. Individual class symbols can be changed by double-clicking on the **Symbol** column of the class list.

5. Choose the **Color ramp** to apply to the classes. Individual class colors can be changed by double-clicking on the **Symbol** column of the class list.

The **Legend Format** field sets the format for all labels. Anything can be typed in the textbox. The lower boundary of the class will be inserted where %1 is typed in the textbox, and the upper boundary of the class will be inserted where %2 is typed.

If **Link class boundaries** is checked, then the adjacent class boundary values will be automatically changed to be adjacent if any of the class boundaries are manually changed.

Other symbol options, such as transparency, color, and output unit, are available by right-clicking on a category row. Advanced settings are available by clicking on the **Advanced** button.

The **Histogram** tab (shown in the following screenshot) allows you to visualize the distribution of the values in the selected column. It is often useful to view the histogram before you decide on the classification method to gain an overview of the distribution of the data and to quickly identify any outliers, skew, large gaps, or other characteristics that may affect your classification choice.

To view the data in the histogram, click on the **Load values** button. You can change the number of bars in the histogram by modifying the value in the **Histogram bins** box. You can view the mean and standard deviation by checking the **Show mean value** and **Show standard deviation** boxes, respectively.

Rule-based vector styling

The rule-based vector style applies one symbol per created rule and can apply maximum and minimum scales to toggle symbol visibility. This vector style is best when you want a different symbol that is based on different expressions or when you want to display different symbols for the same layer at different map scales. For example, if you are styling roads, a rule could be set to make roads appear as thin lines when zoomed out, but when zoomed in, the thin lines will disappear and be replaced by thicker lines that are more scale appropriate.

There are no default values for rule-based styling; however, if a style was previously set using a different styling type, the style will be converted into rule-based when this style type is selected. The following screenshot shows the **Categorized** style type from the previous section that is converted into the **Rule-based** style type parameters for point vector data of surveying levels:

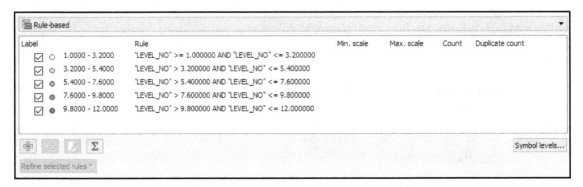

The **Rule-based** style properties window shows a list of current rules with the following columns:

- **Label**: The symbol and label that will be visible in the **Layers** panel are displayed here. The checkbox toggles rule activation; unchecked rules will not be displayed.
- **Rule**: This displays the filter that was applied to the vector dataset to select a subset of records.
- **Min. Scale**: This displays the smallest (zoomed-out) scale at which the rule will be visible.
- **Max. Scale**: This displays the largest (zoomed-in) scale at which the rule will be visible.
- **Count**: This displays the number of features that are included in this rule. This is calculated when the **Count features** button is clicked.
- **Duplicate count**: This displays the number of features that are included in the current and other rules. This is helpful when you are trying to achieve mutually exclusive rules and need to determine where duplicates exist. This is calculated when the **Count features** button is clicked.

To add a new rule, click on the **Add rule** button (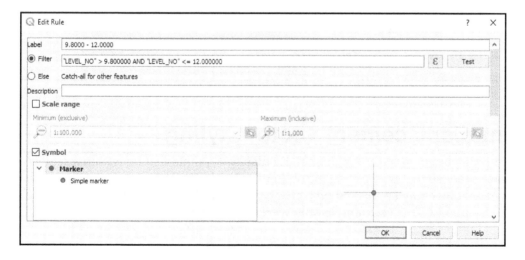) to open the **Rule properties** window. To edit a rule, select the rule and then click on the **Edit rule** button ( ) to open the **Rule properties** window. To remove a rule, select the rule and then click on the **Remove rule** button ( ).

Additional scales, categories, and ranges can be added to each rule by clicking on the **Refine current rules** button. To calculate the number of features that are included in each rule and to calculate the duplicate feature count, click on the **Count features** button.

When a rule is added or edited, the **Rule properties** window (which is shown in the following screenshot) displays five rule parameters, which are as follows:

- **Label**: This should have the rule label that will be displayed in the **Layers** panel.
- **Filter**: This will have the expression that will select a subset of features to include in the rule.
- Click on the ellipsis button to open the **Expression string builder** window. Then, click on the **Test** button to check the validity of the expression.
- **Description**: This has a user-friendly description of the rule.
- **Scale range**: This has the **Minimum (exclusive)** and **Maximum (inclusive)** scales between which the rule will be visible.
- **Symbol**: This has the symbol that will be used to symbolize features that are included in the rule.

None of the parameters are required (**Label**, **Filter**, and **Description** could be left blank); to exclude **Scale range** and **Symbol** from the rule, uncheck the boxes next to these parameters:

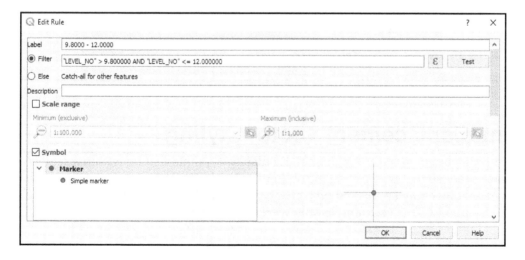

As an example, using the `Populated Places.shp` sample data, capital cities, megacities, and all other places can be styled differently by using rule-based styling. Additionally, each rule is visible to the minimum scale of *1:1*, although they become invisible at different maximum scales.

The following screenshot shows the rules that were created and a sample map of Europe and North Africa:

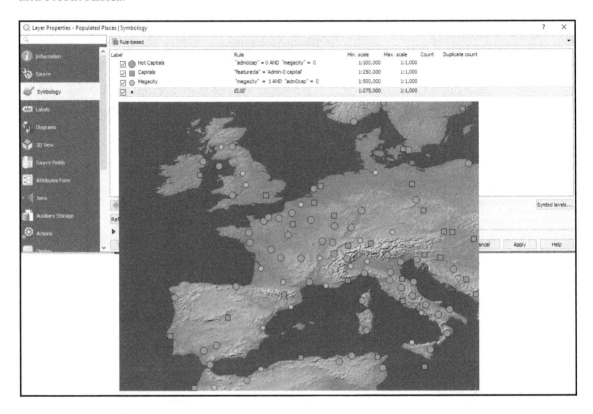

Point-displacement vector styling

The point-displacement vector style radially displaces points that lie within a set distance from each other so that they can be individually visualized. This vector style works best on data where points may be stacked on top of each other, thereby making it hard to see each point individually. This vector style only works with the point vector geometry type.

The following screenshot shows how the **Point displacement** style works by using the **Single symbol** renderer, which is applied to the `Stacked Points.shp` sample data. Each point within the **Point distance tolerance** value of at least one other point is displaced at a distance of the **Circle radius modification** value around a newly created center symbol.

In this example, three groups of circles have been displaced around a center symbol:

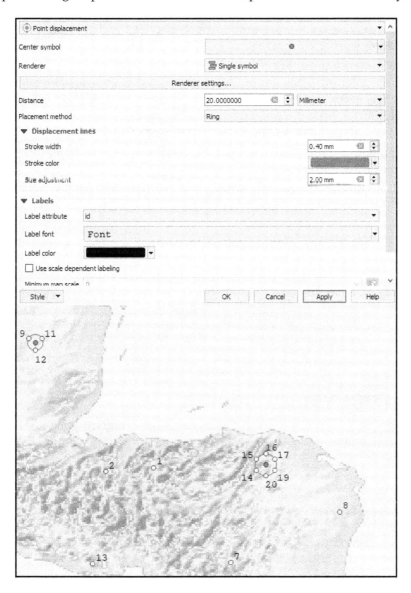

The **Point displacement** style parameters as shown in the preceding screenshot, provide multiple parameters to displace the points, set the sub-renderer, style the center symbol, and label the displaced points. Let's review the parameters that are unique to the point-displacement style:

- **Center symbol**: This contains the style for the center symbol that is created at the location from where the point symbols are being displaced.
- **Renderer**: This contains the renderer that styles the displaced points. Click on the **Renderer settings** button to access renderer settings.
- **Point distance tolerance**: For each point, if another point (or points) is within this distance tolerance, then all of the points will be displaced. The tolerance can be specified in millimeters, pixels, or map units. If choosing map units, further configuration becomes available, allowing for displacement to only happen between certain scale ranges or at certain size ranges.
- **Placement method**: This sets the method of placing the displaced points as either **Ring** or **Concentric rings**.
- **Outline width**: This sets the outline pen width in millimeters, which visualizes the displacement ring value.
- **Outline color**: This contains the outline pen color for the displacement ring.
- **Ring size adjustment**: The number of *additional* millimeters that the points are displaced from the center symbol.

The **Labels** parameters applies to all points (displaced or not) in the vector data. It is important to use these label parameters, rather than the label parameters on the **Labels** tab of the **Layer Properties** window, because the labels set in the **Labels** tab will label the center symbol and not the displaced points.

Point cluster vector styling

Point clustering was introduced in QGIS 3.0. Point cluster rendering is widely used in web mapping applications to condense multiple overlapping or nearby points into a single rendered marker.

Inverted polygons vector styling

The inverted polygons vector style inverts the area that a polygon covers. This vector style only works with the polygon vector geometry type.

The following screenshot shows the **Inverted polygons** style for a polygon of the country of Nigeria on the left and all countries underneath the transparent inverted polygon of Nigeria on the right. Notice that the entire canvas is covered by the inverted polygon, which has the effect of cutting out Nigeria from the map:

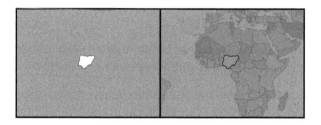

The **Inverted polygons** style parameters rely on a sub-renderer to determine the symbol that was used for the inverted polygons. By choosing the **Sub renderer**, the polygon rendering is inverted to cover the entire map canvas. The following screenshot shows the **Inverted polygons** style parameters that created the inverted polygon of Nigeria:

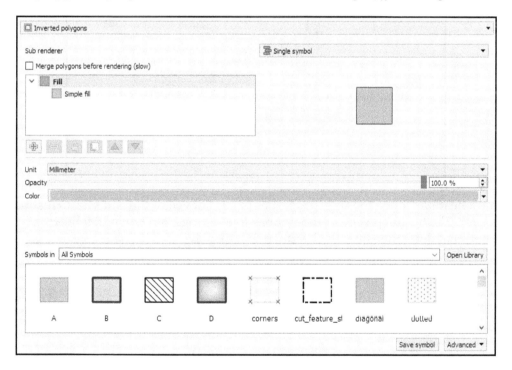

If multiple polygons are going to be inverted and the polygons overlap, **Merge polygons before rendering (slow)** can be checked so that the inverted polygons do not cover the area of overlap.

Heatmap vector styling

A heatmap represents the spatial density of points that are visualized across a color ramp. The heatmap vector style can only be applied to point or multipoint vector geometry types.

As an example, the following screenshot uses a heatmap on a city population point layer to visualize the density and distribution of population in the United States of America:

To create a heatmap, open the style properties of a point vector file and select **Heatmap** as the renderer. The heatmap vector renderer has five settings (as shown in following screenshot) that determine how the heatmap will display:

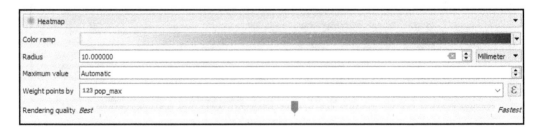

Color ramp sets the color ramp that will be applied across the heatmap. The colors on the left-hand side of the color ramp will represent low density, while the colors on the right-hand side of the color ramp will represent high density.

To reverse this, click on the color ramp. In the **Select Color Ramp** window, click on **Color 1** and change this value to **Transparent**:

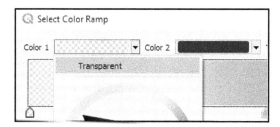

The following is an example of a heatmap rendering for population by cities:

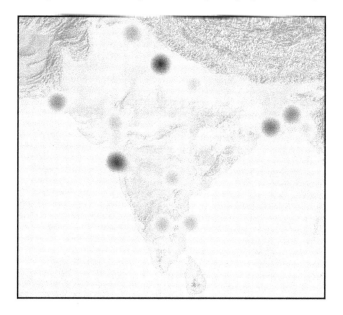

Rarely is the default heatmap satisfactory, so it is recommended to iterate different values of the radius to create the most impactful, and meaningful, heatmap.

Heatmaps are a useful vector layer style when you want to show density and distribution without being bound to geographic or political areas, in contrast to graduated polygon renderers that generate a choropleth map and assume equal distribution of values across the containing area.

2.5 D vector styling

The 2.5 D vector style extrudes two-dimensional (flat) polygons to make them look like they are three-dimensional at a set view angle. This view style is also commonly known as a two thirds (2/3) view, perspective view, or two-and-one-half (2.5) view. Regardless of the name, the 2.5 D vector style can create some compelling scenes. As an example, the following screenshot shows a portion of a city where the buildings are styled using the 2.5 D vector style:

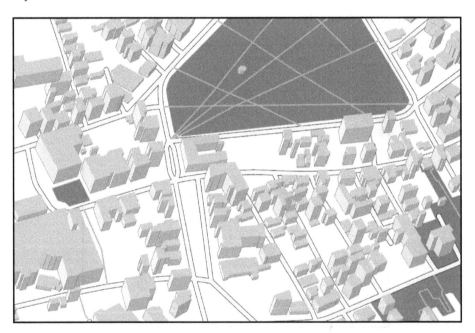

We will now explain the settings that are used to create this 2.5 D city map. First, a polygon vector layer must be added to the map canvas; in our case, this was a building polygon layer. Next, open the style properties and choose the **2.5 D** renderer.

The following screenshot shows the settings that are available for the **2.5 D** renderer:

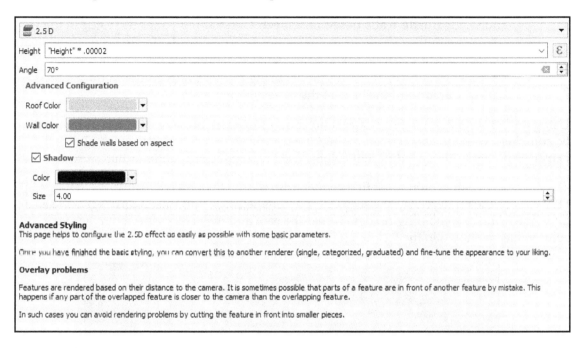

There are a number of settings available for the **2.5 D** renderer. We will now examine each in detail:

- **Height**: This setting determines the height the polygons will be extruded at in map units. This can be either a number you enter, a field in the attribute table, or a calculation. If you enter a number, then all polygons will be extruded to this height. If you enter an attribute field, then the polygons will individually extrude to their associated value. If you click on the **Expression** button (ε), then you can create an expression to be evaluated for the polygons' extrusion height. Often, the height values stored in attribute fields are in a different unit than the map unit and will need to be converted so that the rendering is correct. For instance, if the map units are in meters, but the height attribute is in feet, then you would need to multiply the height attribute by 0.3048.
- **Angle**: This is the angle at which the extruded polygons will lean. The angle can be specified in values between 0 degrees and 359 degrees, with the default being 70 degrees. 0 degrees is the right of the screen and increases in a counter-clockwise direction, so that 90 degrees is the top of your screen.

- **Advanced Configuration**: This section of settings allows you to set the **Roof Color** (top of the extruded polygon), **Wall Color** (sides of the extruded polygon), and toggle **Shade walls based on aspect**. The shading varies the value of the wall colors to simulate shade from a light source. Turning off shading causes the walls to be rendered in the chosen, single, flat **Wall Color**.
- **Shadow**: If this is enabled, a shadow using the set **Color** and with the set **Size** will appear around the base of the extruded polygons.

Using the 2.5 D renderer is pretty easy and straightforward, but a neat feature is that it can be combined with another renderer (single, categorized, or graduated), which allows the extruded polygons to have different colors based on the second renderer that's used.

To combine renderers, you must first render with the 2.5 D renderer, **Apply** the renderer, then change the renderer, set the second renderer's settings, and then **Apply** again. The only downside to combining renders is that the roof and wall colors are set to the same color value, as shown in the following image:

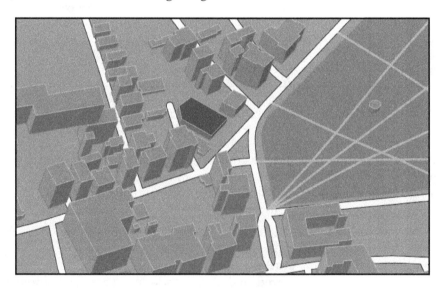

The eight vector layer styles really provide a great deal of control over the way your data is displayed on the map canvas. A number of vector layer styles have been added in recent releases of QGIS and have really increased the cartographic capabilities of the software. Speaking of great improvements to cartographic capabilities, the next section covers layer rendering, which really allows for some great visual effects to be added to your map!

Vector layer rendering

Layer rendering modifies the properties of the vector to change the way it displays and interacts with the layer below and the features within the vector. Layer rendering is a part of vector style properties for all style types and works in the same way, regardless of the selected style type.

In this section, we will discuss the parameters that are available for change in the **Layer rendering** section of vector style properties.

When a vector is first loaded, the **Layer rendering** parameters are set to their default values, as shown in the following screenshot:

The **Layer rendering** section has five parameters – **Layer transparency**, **Layer blending mode**, **Feature blending mode**, **Draw effects**, and **Control feature rendering order** - which will now be explained in detail.

Layer transparency

Layer transparency sets the percentage of transparency for the layer. The higher the transparency value, the more the layers below will be visible through this layer. The transparency is set between the values of 0 (opaque) and 100 (fully transparent).

Layer blending mode

Layer blending mode applies a blending method to the vector that mixes with layers below in the **Layers** panel. A number of blending modes are available to choose from and are commonly found in graphics editing programs. In fact, there are 13 blending modes. Each of these blending modes is discussed in more detail in the *Raster color rendering* section of this chapter.

Feature blending mode

Feature blending mode applies a blending method to the vector that mixes with other features in the same vector layer. A number of blending modes are available to choose from and are commonly found in graphics editing programs. There are 13 blending modes. Each of these blending modes is discussed in more detail in the *Raster color rendering* section of this chapter.

The new blending modes should be explored before you use transparency for overlays. Let's consider an example where we want to add a hillshade to our map to give it some depth. In the following screenshot, the top-left map shows Africa's countries by using polygons and **Normal Layer** blending. A common way to place a hillshade behind polygons is to make the polygons semi-transparent so that the hillshade can give depth to the polygons. However, this tends to wash out the colors in the polygons and the hillshade is muted, which is illustrated at the top-right corner of the following screenshot. Instead, the **Hard Light** or **Multiply Layer** blending methods (illustrated at the bottom-left and bottom-right corners in the following screenshot, respectively) can be used to maintain strong color and include the hillshade:

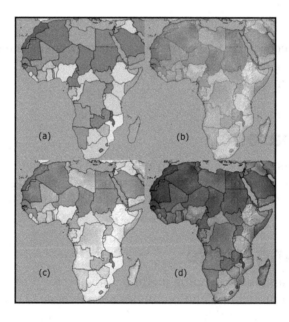

Control feature rendering order

If enabled, **Control feature rendering order** allows you to define the order in which attributes will be rendered. The rendering order is set using expressions that define subsets of attributes for a rendering group, whether than subset should render in ascending or descending order, and whether **NULL** values should render first or last.

The following screenshot shows two expressions and their rendering options. In this case, the features that were selected in row 1 will render on top of the features that were selected in row 2, and within each row, they will be rendered in either ascending or descending order. The **NULL** values will be rendered first or last:

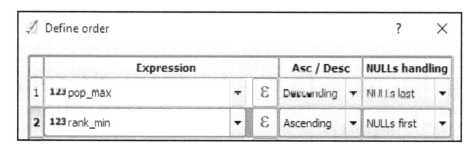

For our example, let's make sure that the larger proportional circles are rendered below smaller circles so that the smaller circles will not be hidden behind the larger circles. The following screenshot shows default rendering on the left (larger circles rendered on top of smaller circles) and controlled rendering on the right (smaller circles on top of larger circles):

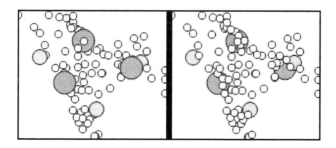

To achieve this rendering of larger circles on the bottom, we simply set the attribute controlling the size of the circles to render in **Descending** order, as shown in the following screenshot:

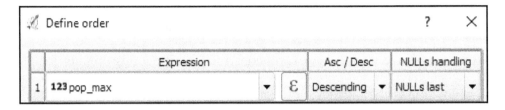

Layer rendering can really improve the look of your map. So, experiment with the layer rendering methods to find the ones that work best for your overlays.

Using diagrams to display thematic data

To add a diagram, open the vector's **Layer Properties** window and then click on the **Diagrams** tab. The **Diagrams** tab, as shown in the following screenshot, has five sections, as well as a **Diagram type** selector.

 QGIS supports the addition of three diagram types as overlays on top of vector data. The three diagram types are *pie chart*, *text diagram*, and *histogram*. The underlying vector data can still be styled to provide a nice base map.

The **Diagram type** selector allows you choose which type of diagram to use: **Pie chart**, **Text diagram**, or **Histogram**. When you choose a type of diagram, the other five sections change to contain the options that are appropriate for the diagram type.

The sections are **Attributes, Rendering, Size, Placement, Options**, and **Legend**. **Attributes** is common to all diagram types and provides the mechanism for adding attributes to diagrams. **Size** and **Placement** contain parameters that are shared by all of the diagram types. **Appearance** and **Options**, however, contain parameters that are unique to each diagram type:

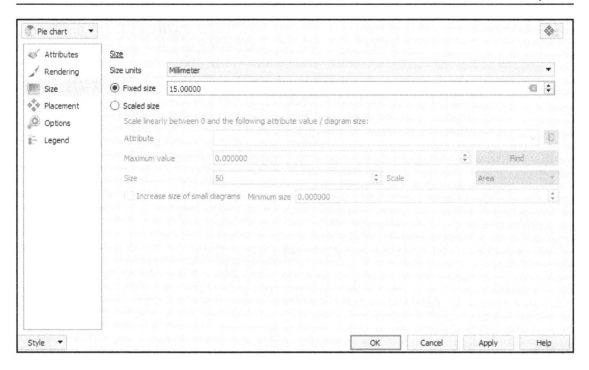

Parameters that are common to all diagram types

Since the **Size**, **Placement**, and **Attributes** sections are common to all diagrams, the following section will cover these parameters first.

The following three sections will cover the parameters that are unique to each type of diagram.

Diagram size parameters

The **Size** section, which is shown in the following screenshot, provides the following parameters:

- **Size units**: This contains the unit of the **Fixed size** parameter, which can be set to either millimeters (**mm**) or **Map units**.
- **Fixed size**: If this is selected, all charts will have the specified area or diameter (for a pie and text chart) or length (for a bar chart).

- **Scaled size**: If **Fixed size** is *not* selected, then the length (for a bar chart), area, or diameter (for a pie and text chart) set by the **Scale** parameter of the charts will be scaled down linearly from the selected **Size** value. The selected **Size** value represents the maximum or set **Attribute** value. To enable QGIS to determine the maximum attribute value, click on the **Find** button.
- **Increase size of small diagrams**: If checked, this parameter sets the **Minimum size** value to which the charts will be scaled:

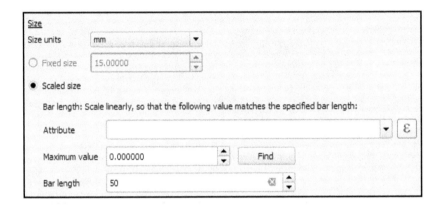

Diagram placement parameters

The **Placement** section, as shown in the following screenshot, provides the following parameters:

- **Placement**: This sets the placement of the chart. The available options are **Around Centroid**, **Over Centroid**, **Perimeter**, and **Inside Polygon**.
- **Distance**: The distance the diagram will be offset from the centroid of the related feature. This is available when **Around Centroid** is selected for the **Placement** parameter.
- **Data defined position**: If this is checked, the x and y positions of the chart can be set by attributes.
- **Priority**: This sets the priority defining how likely it is that the diagram will be rendered in relation to other objects to be rendered on the map canvas. Objects or labels with higher priorities will be rendered first. Lower priority objects or labels may be left off of the map if there is not enough room.
- **Diagram z-index**: Sets the z-index for the drawing order of the diagrams in relation to other objects and labels. Objects and labels with higher z-index numbers will be drawn on top of those with lower z-index numbers:

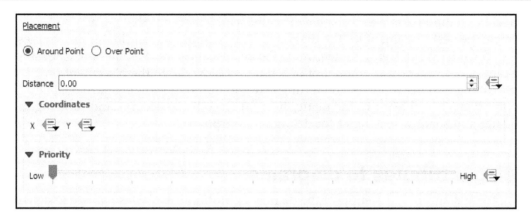

- **Automated placement**: The button in the upper-right corner () opens the
  Automated Placement Engine window (shown in the following screenshot). This
  provides more parameters to fine-tune the placement of charts, such as showing
  all charts and showing partial labels:

The **Search method** parameter lists five different search methods that are available to determine the best placement for the diagrams. The **Number of candidates** determines how much to consider when trying to determine the best location for the diagram. Choosing more candidates will provide more options, but may slow down the placement process. The remaining options allow for converting text into outlines, showing partial labels, showing all labels and features (no matter how much collision occurs), and showing candidates and shadow rectangles for debugging purposes.

Adding attributes to diagrams

Each of these diagram types support the display of multiple attributes that are assigned in the **Attributes** section. To add or remove attributes, you must move (or build) an expression from the **Available attributes** list to the **Assigned attributes** list. Attributes in the **Assigned attributes** list will be used in the diagram.

There are two ways to add an attribute to the **Assigned attributes** list, which are as follows:

- Select the attribute(s) from the **Assigned attributes** list. Then, click on the **Add attribute** button (⊕).
- Click on the **Add expression** button (ε...) and then create an expression that will be added as a single entry.

Once an attribute has been added, the **Assigned attributes** colors can be changed by double-clicking on the color patches in the **Color** column. Additionally, the **Legend** (that is, label) for each attribute can be changed by double-clicking on the entry in the **Legend** column.

The following screenshot shows an example of added attributes, with their colors and legends set:

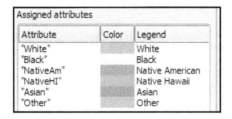

Creating a pie chart diagram

A pie chart diagram displays attributes in a round pie chart, where each attribute occupies a pie slice proportional to the percentage that the attribute represents from the total of all attributes added to the pie chart. As an example, the following screenshot shows a portion of a state, with pie charts showing the proportion of different racial populations in each county:

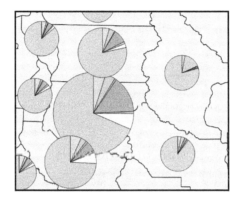

By reviewing the pie charts in the preceding screenshot, let's note down a few things:

- The pie slices differ for each county's race attributes and are colored based on each attribute's selected color
- The sizes of the pie charts vary based on the total population in each county
- The pie slice outlines are black and thin
- The pie charts are displayed above the centroid of each polygon

These four noted items, among others that are not noted, are all customizable using the parameters that are available in the **Diagrams** tab of the **Layer Properties** window. The **Diagrams** tab for pie charts has three sections with unique parameters: **Rendering**, **Size**, and **Placement**.

The Rendering section, which is shown in the following screenshot, provides the following parameters:

- **Transparency**: This is used to specify the transparency percentage for the pie chart
- **Line color**: This contains the color of the lines surrounding the pie and in-between pie slices.

- **Line width**: This contains the width of the lines surrounding the pie and in-between pie slices.
- **Start angle**: This is used to specify the angle from which the pie slices will begin to rotate in a clockwise manner. The available options are **Top**, **Right**, **Bottom**, and **Left**.
- **Show all diagrams**: If this is checked, all pie charts are displayed. If unchecked, pie charts that are not able to be placed without colliding with other pie charts will not be placed.
- **Scale dependent visibility**: If this is checked, the **Minimum** and **Maximum** visibility scales can be set:

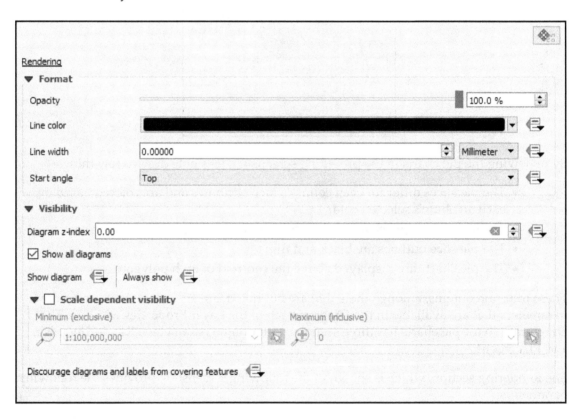

Creating a text diagram

The text diagram displays attributes in a round circle, where each attribute occupies a horizontal slice of the circle and the attribute value is labeled inside the slice. As an example, the following screenshot shows a portion of a state's counties, with text diagrams showing the Hispanic population in the top half and the non-Hispanic population in the bottom half of the circle:

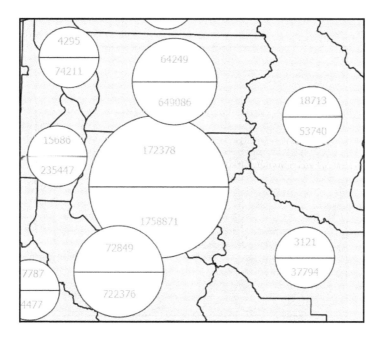

By reviewing the text diagrams in the preceding screenshot, let's note down a few things:

- The labels report each race's attribute and are colored based on the selected attribute colors
- The size of the diagram varies according to the total population of each county
- The horizontal slice outlines are black and thin
- The text diagrams are displayed over the centroid of each polygon

These four noted items, among others that are not noted, are all customizable using parameters that are available in the **Diagrams** tab of the **Layer Properties** window. The **Diagrams** tab for text diagrams has one section with unique parameters: **Appearance**.

Creating a histogram chart diagram

The histogram chart diagram displays attributes in a histogram/bar chart, where each attribute can be visualized as a bar that varies in length in proportion to the attributes' values.

As an example, the following screenshot shows a portion of a state's counties with histogram charts showing the Hispanic population as one bar and non-Hispanic population as the other bar:

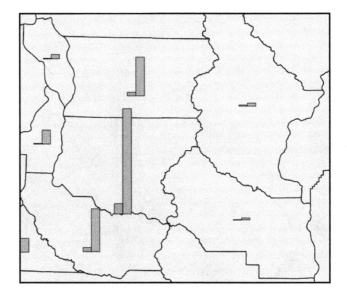

By reviewing the histogram charts in the preceding screenshot, let's note down a few things:

- The bars report each race's attribute and are colored according to the selected attribute colors
- The bar outlines are black and thin
- The histogram charts are displayed over the centroid of each polygon

These three noted items, among others that are not noted, are all customizable using parameters that are available in the **Diagrams** tab of the **Layer Properties** window. The **Diagrams** tab for text charts has one section with parameters that are unique to the histogram diagram type: **Options**.

The **Options** section, which is shown in the following screenshot, provides the **Bar Orientation** parameter. This parameter sets the orientation of the bars. The options that are available are **Up**, **Down**, **Right**, and **Left**:

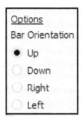

Saving, loading, and setting default styles

Now that you have set the styles that you want for raster and vector layers, you will likely want to save the styles so that they can be used again later or applied to other layers. The **Style** button, which is shown in the following screenshot, is always displayed at the bottom of the **Layer Properties** window:

In this section, we will use this button to save a style to a style file, load a saved style file, set and restore a default style, and add and rename styles to the current QGIS project.

Saving a style

QGIS can save styles in two file formats: QGIS Layer Style File (.qml) and SLD File (.sld). The .qml style file is specific to QGIS, while the .sld style file is usable by other programs to style files. In general, you should plan on saving styles using the .qml file type as it does the best job of saving your styles; however, if portability is a priority, then the .sld file is the better choice. To save a style, open the **Layer Properties** window, set the style that you wish to save, then go to **Style** | **Save Style** and save the style as either a .qml or a .sld file. The saved style file can later be loaded and applied to other data files.

> To have a style always apply to the layer you are saving the style from, save the style file as a .qml and give it the same name as the source file. For example, if the shapefile name was Coastlines.shp, then you should save the style file as Coastlines.qml. Creating a style file with the same name has the same effect as setting the default style.

Loading a style

QGIS can load styles from two formats: QGIS Layer Style File (.qml) and SLD File (.sld). To load a style, open the **Layer Properties** window, click on the **Style** | **Load Style** button, and open the style file that you wish to load. This will apply the style to the layer. QGIS will try and apply the styles, even if the geometry of the style is not applicable to the geometry of the current layer. This can create unexpected output, so it is usually best to load styles that match the geometry type.

Setting and restoring a default style

To set a layer's current style as the default style for use in other QGIS projects, go to **Style** | **Save As Default** in the **Layer Properties** window. This will save a .qml file with the same name as the layer on disk. When the layer is added to the map canvas (in this or another QGIS project), the saved default style will be automatically applied.

If you have made changes to the style of a layer that had a default style and wish to revert back to the default style, go to **Style** | **Restore Default Style**.

Lastly, to remove a default style for a layer, delete the .qml file of the same name as the layer on disk.

Adding and renaming styles in the current QGIS project

Styles can be saved inside QGIS project files (.qgs). In fact, all .qgs files already contain one style named default, which stores all of the style choices you make in the QGIS project. Styles that are added in this way will only exist when the .qgs project is loaded, which can be useful when you want to store a set of styles that are specific to a project inside the .qgs file instead of relying on separate style files.

To add the currently set style options as a new style inside the .qgs file, open the **Layer Properties** window, go to **Style | Add**, and provide a name for the style in the **New style** dialog box. Once added, the new style will appear and be selectable at the bottom of the **Style** button. Only one style can be active at a time, so checking one style will uncheck all other styles.

To rename a style that's stored in a .qgs file, open the **Layer Properties** window, make sure that the style you wish to rename is the currently selected style, and then go to **Style | Rename Current**.

Summary

In this chapter, we looked at the steps to style vector and raster data in QGIS. We first covered how to pick colors using the new color picker. Then, we covered how to create and manage color ramps using the style manager. Next, we reviewed the different ways to style singleband and multiband rasters, create a raster composite, as well as how to overlay rasters using renderers. Vector styling was reviewed next and we covered the six different style types. We also looked at how to use vector renderers for layer overlays. Next, we toured the three diagram types that can be visualized on top of vector datasets. We finished this chapter with instructions on how to save and load styles for use in other QGIS projects.

In the next chapter, we will move from viewing data to preparing data for processing. Preparation topics will range from spatial and aspatial queries and converted geometry types to defining new coordinate reference systems.

Creating Spatial Databases

3

This chapter covers the creation and editing of spatial databases. While there are many spatial database types that QGIS can connect to, such as Oracle SDO, MySQL, SQL Server, and of course PostGIS, for ease of use, we're going to concentrate on a spatial database type that requires no installation and that can be easily shared. SpatiaLite (`.sqlite`) and GeoPackage (`.gpkg`) are the most common portable databases.

GeoPackage is being pushed as the replacement for the Industry Standard GIS file type, the SHP file, and you may notice as you start to use QGIS more that it crops up as the first choice of file type to open or export to.

In this chapter, we will touch on the core concepts of databases but moreover explore the following topics:

- Fundamental database concepts
- Creating a spatial database (GeoPackage)
- Connecting to a GeoPackage
- Importing data into a GeoPackage
- Working with tables
- Exporting tables out of a GeoPackage
- Creating queries and views

 As of QGIS 3.6, it is not possible to import rasters into a GeoPackage, just vectors. We are not going to cover how to import rasters into GeoPackages in this book. However, if you are interested in this, we suggest looking at the GDAL library for further reference (`https://www.gdal.org/drv_geopackage_raster.html`).

Fundamental database concepts

A **database** is a structured collection of data that provides multiple benefits over data that's stored in a flat file format, such as a shapefile or KML. These benefits include complex queries, complex relationships, scalability, security, data integrity, and transactions, to name a few. Using databases to store geospatial data is relatively easy, considering the aforementioned benefits.

 There are multiple types of database; however, the most common type of database, and the type of database that this chapter will cover, is the relational database.

Describing database tables

A relational database stores data in tables. A **table** is composed of rows and columns, where each row is a single data record and each column stores a field value associated with each record. A table can have any number of records; however, each field is uniquely named and stores a specific type of data.

A data type restricts the information that can be stored in a field, and it is very important that an appropriate data type, and its associated parameters, be selected for each field in a table. The common data types are as follows:

- Integers (whole numbers), for example, 9, 27, and 345
- Floats (with decimal points, sometimes called real numbers), for example, 1.5 and 4.75
- Text, for example, abc and abc 123
- Boolean, for example, 'True' or 'False'
- Date, for example, 'dd/MM/yyyy'

Each of these data types can have additional constraints set, such as setting a default value, restricting the field size, or prohibiting null values.

In addition to the common data types we mentioned previously, some databases support the geometry field type, allowing the following geometry types to be stored: Point, LineString, Polygon, MultiPoint, MultiLineString, MultiPolygon and GeometryCollection.

The multi-point/line/polygon types store multi-part geometries so that one record has multiple geometry parts associated with it.

Knowing about table relationships

A table relationship connects records between tables. The benefit of relating tables is reducing data redundancy and increasing data integrity. To relate two tables together, each table must contain fields that have values in common. For example, one table may list all of the parcels in a city, while a second table may list all of the land owners in the city. The parcels table can have an `OwnerID` field that also exists in the owners table so that you can relate the correct owner from the owners table to the correct parcel in the parcels table. This is beneficial in cases where one owner owns multiple parcels, but we only have to store the owner information one time, which saves space and is easier to update as we only have to update owner information in one record.

 The process of organizing tables to reduce redundancy is called **normalization**. Normalization typically involves splitting larger tables into smaller, less redundant tables, followed by defining the relationship between the tables.

A field can be defined as an index. A field that's set as an index must only contain values that are unique for each record, and therefore it can be used to identify each record in a table uniquely. An index is useful for two reasons. Firstly, it allows records to be quickly found during a query if the indexed field is part of the query. Secondly, an index can be set to be a primary key for a table, allowing table relationships to be built.

A **primary key** is one or more fields that uniquely identify a record in its own table. A foreign key is one or more fields that uniquely identify a record in another table. When a relationship is created, the record(s) from one table is linked to the record(s) of another table. With related tables, more complex queries can be executed and redundancy in the database can be reduced.

Reviewing the basics of the structured query language

Structured Query Language (SQL) is a language that was designed to manage databases and the data contained within them. Covering SQL is a large undertaking and is outside the scope of this book, so we will only cover a quick refresher that is relevant to this chapter.

SQL provides functions to select, insert, delete, and update data. Four commonly used SQL data functions are as follows:

- `SELECT`: This retrieves a temporary set of data from one or more tables based on an expression. A basic query is `SELECT <field(s)> FROM <table> WHERE <field> <operator> <value>;`, where `<field>` is the name of the field from which values must be retrieved and `<table>` is the table on which the query must be executed. The `<operator>` part checks for equality (such as =, >=, and `LIKE`), while `<value>` is the value to compare against the field.
- `INSERT`: This inserts new records into a table. The `INSERT INTO <table> (<field1>, <field2>, <field3>) VALUES (<value1>, <value2>, <value3>);` statement inserts three values into their three respective fields, where `<value1>`, `<value2>`, and `<value3>` are stored in `<field1>`, `<field2>`, and `<field3>` of `<table>`.
- `UPDATE`: This modifies an existing record in a table. The `UPDATE <table> SET <field> = <value>;` statement updates one field's value, where `<value>` is stored in `<field>` of `<table>`.
- `DELETE`: This deletes record(s) from a table. The following statement deletes all of the records matching the `WHERE` clause: `DELETE FROM <table> WHERE <field> <operator> <value>;`, where `<table>` is the table to delete records from, `<field>` is the name of the field, `<operator>` checks for equality, and `<value>` is the value to check against the field.

Another SQL feature of interest is view. A **view** is a stored query that is presented as a table, but is actually built dynamically when the view is accessed. To create `VIEW`, simply preface a `SELECT` statement with `CREATE VIEW <view_name> AS`, and a view named `<view_name>` will be created. You can then treat the new view as if it were a table.

Creating a spatial database

Creating a spatial database in QGIS is a simple task to perform. QGIS comes with two spatial database types that can be easily created without the need to install or set up anything.

SpatialLite and GeoPackages are relational databases that are stored as flat files. What this means is that the data is stored within a single file and not on traditional server-based database applications such as Oracle or Postgres/PostGIS.

We will be dealing with GeoPackages in this chapter, which is an OGC compliant GeoSpatial format. It is open source format and is consumed by most GIS packages, including ESRI, MapInfo, and QGIS.

SpatiaLite (**SQLite**) is built on a single-user architecture, which makes the installation and management virtually nonexistent. The trade-off, however, is that it neither does a good job of supporting multiple concurrent connections, nor does it support a client-server architecture. For a more complex DBMS, PostGIS is an excellent open source option.

Click on **Layer** | **CreateLayer** | **New GeoPackage Layer**. Alternatively, in the **Browser** panel, right-click on **GeoPackage** and choose **Create Database**. Both options bring you to the same window:

1. In the **New GeoPackage Layer** window, click on the **Browse** button at the end of the label database and browse to a location where you want to store your **GeoPackage**
2. Give the **GeoPackage** a database name, for example, New_Sites
3. Click on **OK** and you will come back to the **New GeoPackage Layer** window
4. Set the **Table name** and set the value to proposed_sites

When creating a new GeoPackage/Database, QGIS will automatically give the **Table name** as that of the name of the GeoPackage. So, be sure to change the name of the table otherwise you might get confused. It is a good database practice not to give databases, schema, or table names the same name.

5. Set the **Geometry type** to **Polygon**

We will apply good database conventions by replacing all spaces between words with underscores. This is generally seen as a best practice for table names in databases. Furthermore, it is not advisable to start tables with numbers. The simple reason for this is that it makes writing SQL statements and generally managing your data within a database a lot easier.

6. Set the **EPSG** to **27700 'British National Grid'** (if you are doing the follow along exercises): To do this, click on the globe icon at the end of the drop-down menu and filter on **27700**

7. In a new field, create the following columns and leave these as `Text data` types:
    - `site_name`
    - `reference_number`
    - `land_release_date`
    - `proposed_construction_date`
    - `proposal_type`

8. Click on **OK** to create the **GeoPackage**

We'll revisit `New_Sites.gpkg` later in this chapter.

Connecting to a GeoPackage

To be able to load data into and to see what data is stored within a GeoPackage, you first need to make a connection to it.

In the **Browser panel** tab, right-click on **GeoPackage** and choose **New Connection**. Navigate to where the `new_sites.gpkg` GeoPackage is stored.

Once connected, you can expand the GeoPackage and see the layers and tables stored within it.

Importing data into a GeoPackage

Let's see how we can import data into a GeoPackage. We will see how we can import a vector file and layers from the map canvas in this section.

Importing a vector file

Make sure that you've made a connection to the GeoPackage that you want to load data into (see the *Connecting to a GeoPackage* section).

QGIS can consume a lot of file types in a GeoPackage, beyond those of just SHP, TAB, or KML. Irrespective of the supported file type that you want to import, the process is the same:

1. Click on **Database | DBManager** and expand the **GeoPackage** section so that you can see the `new_sites` GeoPackage.
2. Click on the **GeoPackage** so that it is highlighted (in blue).
3. Click on the **Import Layer/File** icon in the toolbar.
4. In the **Import Vector Layer** window, click on the **Browse** button at the end of the **Input Field**. Navigate to the sample data for this chapter and choose `proposed_sites_points.shp`.
5. Click on the **Update Option** button and leave all of the settings as they are. Then click on **OK**.

The data should have successfully been imported into the `new_site` GeoPackage.

Close the DBManager and go to the **Browser** panel. You should now see the `proposed_sites_points` table in the `new_site` GeoPackage. If you can't, click on the refresh button at the top of the panel.

Importing a layer from map canvas

If you have a layer in the map canvas that you want to load into a GeoPackage, this can be done through the DBManager. This feature is great if you have loaded data in from an external web source such as a WFS.

Let's follow these steps:

1. Load `proposed_sites_points.shp` onto the new map canvas from the sample data. As we covered in earlier chapters, this can be done through the **Data Source Manager**. Alternatively, in your desktop file explorer, navigate to the exercise material, drag `proposed_sites_points.shp`, and drop it onto the map canvas.
2. Click on **Database | DBManager** and choose the `new_site` GeoPackage.
3. Click on **Import Layer/Filter**. In the **Input** field, you will see the `proposed_sites_points` layer.
4. Click on **Update Options**.
5. In the table field, change the table name to `proposed_sites_points_2`. Leave all of the other options as they are and click on **OK**.

The data should have successfully been imported into the `new_site` GeoPackage. Close the **DBManager** and go to the **Browser** panel. You should now see the `proposed_sites_points_2` table in the `new_site` GeoPackage. If you can't, click on the refresh button at the top of the panel.

Let's explore the **Options** prior to a new vector layer being imported into a GeoPackage:

- Primary key
- Geometry column
- Source SRID
- Target SRID
- Replace destination table (if it exists)
- Create single-part geometries instead of multi-part
- Convert field names into lowercase
- Create spatial index

Working with tables

Excited about actually working on databases? Well, let's start off with exploring how to work with a table.

Creating a new table within an existing GeoPackage

There are chiefly two ways to create new tables within an existing GeoPackage in QGIS, that is via the **Browser** panel and via the DBManager:

1. In the **Browser** panel, find the GeoPackage that you want to create a new table/layer for
2. Right-click on it and then choose **Create a new table or layer**
3. Go through the process of giving your new table a table name, a geometry type, and fields

When discussing databases, the term table is used to specify data that has geometry (point, lines, and polygons, for example) but also data that is simply a table without any geometry. Both GeoPackages and SpatiaLite contain both table types. You may have noticed that, when you create a new GeoPackage, you are asked about the type of table you are creating—**No Geometry**, **Point**, **Line**, and so on.

You will note that you can't change the database name/location or the EPSG. These were set when the database was originally created. If you want to create a new layer that has a different projection to the GeoPackage, then you will to create the data first as a vector layer (SHP) and then import that data into the GeoPackage via the DBManager.

Renaming a table

Unlike SQLite databases that are maintained in QGIS, GeoPackages have the ability to have tables names changed. This can't be done via the **Browser** panel - this needs to be done via the DBManager:

1. To rename a table/layer, open **DBManager** (**Database | DBManger**) and expand the GeoPackage that contains the table that you want to rename
2. Right-click on the layer and choose a new name

 Note that QGIS will not store the name change if the table/layer is open in an existing project or if the GeoPackage is being accessed in the **Browser** panel.

Editing an existing table field/column

As with the renaming of tables/layers, you need to make sure that no one has an active connection to the GeoPackage. Where GeoPackages are being shared across networks or via `http/s` requests, it might prove hard to make such changes.

In the DBManager, expand the GeoPackage that you want to change:

1. Right-click on the table and choose **Edit.**
2. In the table **Properties** window, you can now edit selected columns.

3. Change the columns as required. If you are doing the follow-along exercises, you will see that the `proposed_sites_points` table had a column that is called `ref_number`. So that we can later join this field with the `proposed_site` polygon table, we want to rename the `ref_number` field to `reference_number`. We can retain all of the other property types.
4. Click **OK**.

Deleting a GeoPackage table

Like with creating a table, there are also two ways to delete an existing table in a GeoPackage: via the DBManager or via the **Browser** panel. You will need to have created a **New Connection** to the GeoPackage before you can delete a table (see the *Connecting to a GeoPackage* section).

In the **Browser** panel, go to the GeoPackage that contains the table that you want to delete. Expand the GeoPackage so that you can see all of the tables within it.

In the DBManager, expand the GeoPackage that contains the table that you want to delete. Right-click on the table that you want to delete and then choose **Delete**.

Exporting tables out of a GeoPackage

You can export tables from GeoPackages to many different formats such as ESRI Shapefile (`.shp`), dBASE (`.dbf`), text (`.txt`), comma-separated values (`.csv`), Excel spreadsheets (`.xls`/`.xlsx`), AutoCAD DXF (`.dxf`), Geography Markup Language (`.gml`), Keyhole Markup Language (`.kml`), Geometry JavaScript Object Notation (`.geojson`), and GeoRSS (`.xml`). You can also export data into a new GeoPackage.

On the map canvas, right-click on the layer that you want to export:

1. Choose **Export | Save Features As...**
2. Navigate to where you want to save your export, name the file, and set the file type
3. Click **OK**

Creating queries and views

Unfortunately, in QGIS 3.2, the functionality to run SQL queries against GeoPakages has not been fully developed as of yet in the DBManager. To carry on exploring SQL queries, we are going to focus on SpatiaLite. To do this, we first need to connect to a SpatiaLite database, which can be found in your `Chapter 3`, *Creating Spatial Databases*, samples download.

In the **Browser** panel tab, right-click on **SpatiaLite** and choose **New Connection**. Navigate to `open_data.sqlite` from the `Chapter 3`, *Creating Spatial Databases*, samples.

Once connected, you can expand the `open_data` database and see the layers and tables stored within it.

Using select statements

Open the **DBManager** (**Database | DBManager**), navigate to **SpatiaLite** in the tree, find the **open_data** database, and then click on it and expand it so that you can see all of the tables within it. Click on the **SQL Window** icon.

In the new tab called **Query** (`open_data.sqlite`), you can start to perform queries.

Let's start with our first query, which is going to be a simple select statement:

1. In the main window, start to type your select statement. You will see that some words will auto populate, which is a great time-saver:

    ```
    SELECT * FROM SU_Road WHERE classifica = "A Road"
    ```

2. Once the statement is written, click on the **Execute** button. The results of the query can be seen in the lower half of the window.

If you have data that contains a lot of fields and you only want a few returned, then you can structure your SQL in the following way:

```
SELECT distname, roadnumber, classifica as classification, geom FROM
SU_Road WHERE classifica = "A Road"
```

In this example, we have only returned four fields (`distname, roadnumber, classifica,` and `geom`), but we have also given the `classifica` field an alias called `classification` and added a `WHERE` clause.

Using `Like` in a SQL statement is a great way to find results that match a keyword. In the following example, we are trying to find, from the `classifica` field, all of the roads that are like `'A Roads'`.

The `%` sign acts as a wild card and can be placed at the beginning or the end of a keyword search:

```
SELECT * FROM SU_Road WHERE classifica Like "%A Road%"
```

You can start to use all of these clauses together:

```
SELECT distname as road_name, roadnumber, geom FROM SU_Road WHERE
classifica LIKE 'A Road%' AND roadnumber = 'A25'
```

Reconnect or **Refresh** the database when queries that result in new tables are being created.

Creating a spatial view

Creating a `View` button will create a `SELECT` statement with the `geom` column as a line:

```
Create View a_roads_view as SELECT distname, roadnumber, classifica as
classification, geom FROM SU_Road WHERE classifica = "A Road"
```

If you create a `View` button with just an SQL statement, then the line geometry is not created and you are left with only a table. A further step is then required to update the geometry:

```
INSERT INTO geometry_columns (f_table_name, f_geometry_column,
geometry_type, coord_dimension, srid, spatial_index_enabled)
 VALUES ("a_roads_view", "geom", 1005, 3, 27700, 0)
```

Dropping a spatial view

We can use the following code to delete a spatial view:

```
DROP VIEW a25_view
 DELETE FROM views_geometry_columns WHERE view_name = 'a25_view'
```

Summary

I hope you now know the steps to handle databases in QGIS. While QGIS can handle multiple databases, we used the GeoPackage as it provides a good amount of functionality with little overhead or administration.

Using the DBManager, you can perform a number of operations on databases. Some operations of note are creating indices, spatial and aspatial views, importing and exporting, and performing queries. From the introduction to the DBManager and GeoPackage (and SpatiaLite) in this chapter, you are now well-equipped to write more complex queries that take full advantage of SQL commands and SpatiaLite SQL extension commands.

In the next two chapters, we will start to look at styling and rendering options and techniques for vector and raster data. This chapter looks at how to work with vector data especially how to merge vectors, convert vector geometries, how to use basic geoprocessing tools for vector data, define coordinate reference system and so on. In the next chapter, we will cover working with raster dataset, specifically how to merge rasters, convert and clip rasters etc. These chapters will equip you with rudimentary knowledge of working with geospatial data before delving deep into topics like Processing Toolbox, Graphical Modeler, creating QGIS plugins with PyQGIS and PyQGIS scripting—all of which will be covered in the upcoming chapters.

4
Preparing Vector Data for Processing

Typically, raw data that's obtained for a **Geographic Information System** (**GIS**) project needs to be massaged for use in a specific application. It may need to be merged, converted into a different geometry type, saved to the coordinate reference system of the project, subset to the extent of the study area, or subset by attribute values. While QGIS provides a powerful set of tools that can handle many types of vector preparation and transformation tasks, this chapter will cover what we consider to be commonly used vector preparation tasks. Many of the tools that will be covered in this chapter can be found in the **Vector** menu in QGIS; however, others are available in the processing toolbox.

The following topics will be covered in this chapter:

- Merging vectors
- Converting vector geometries
- Adding geometry columns to an attribute table
- Using basic vector Geoprocessing tools
- Defining coordinate reference systems
- Advanced field calculations

Merging vectors

The **Merge Vectors** tool does exactly what it says to does—merge vector files. The merge tool allows for a large range of vector file types to be merged. Typically, however, this will be shapefiles. The input vectors must be in a common coordinate reference system and be of the same geometry type (point, line, or polygon), and should contain common attributes, for example, a reference number or type value. In these cases, the data may need to be merged to form a seamless layer covering the study area.

To access the tool, go to **Vector**, then **Data Management Tools**, and finally **Merge Vector Layers**. In the **Merge Vector Layers** window, set the **Input layers** to be merged by clicking on the browse button at the end of the **Input layers** field:

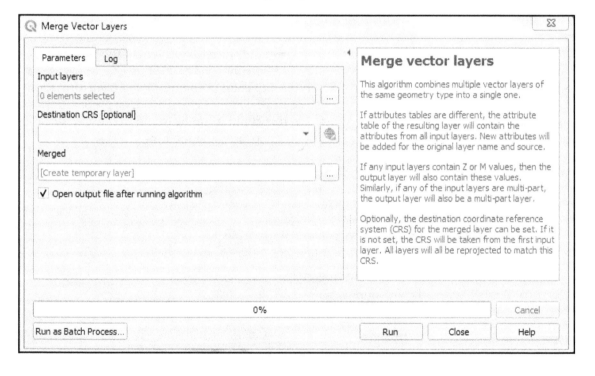

If you have vector layers already loaded in QGIS, these will be displayed in the next window. You can either choose to merge specific layers by individually clicking on them or select all. Alternatively, you can navigate to where the vector data that needs to be merged resides:

Set the **Destination CRS**. This might be the same as the original data to be merged, or it could be different.

In the **Merged** textbox, you can specify to either have the data written to a location with a specified name and file type, or by leaving this section blank, the file will be written as a temporary file into the map canvas, in this instance, called **Merged**. To save the output to file, click on the browse button at the end of the **Merged** textbox and choose to save to file (where you can set the vector file type and give it a name). You can also save to `GeoPackage` or write the data to a `PostGIS` database if you have one set up.

Notes on a temporary file: The data that's written to a temporary file will only last for as long as you have the project open. When you close the project, this data is lost. However, the layer will remain in the project, although there is no data within it.

Click on the **Run in Background** button to start the merge tool. Depending on the size of the task, this could take seconds or minutes. The log tab at the top of the window will show you the processing processes involved in merging the data and indicate whether there are any errors in the task of merging the data.

Converting vector geometries

Sometimes, it is necessary to make conversions among point, line, and polygon vector geometries. For example, you may need to generate point centroids from ZIP code polygons or a town boundary polygon from a line layer. Such conversions may be necessary to put the data into the most appropriate geometry for analysis. For example, if you need to determine the acreage of parcels, but they are provided in a line format, you will need to convert them into polygons to calculate their areas. Sometimes, you may want to convert geometries for cartographic reasons, such as converting polygons into points to create label points. The following tools can be found in **Vectors | Geometry Tools**.

Creating polygon centroids

Let's create polygon **Centroids**—you can see this in the following screenshot:

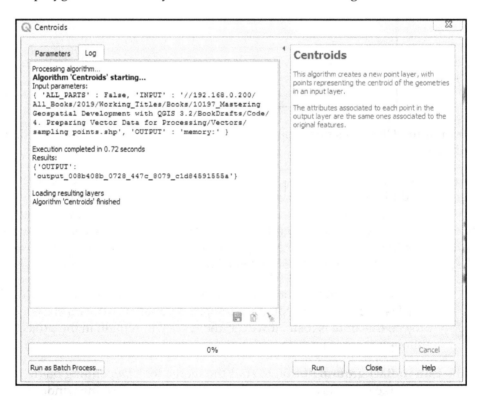

In the **Centroids** window, set the layer to be processed by clicking on the browse button at the end of the **Input layer** field.

If you have vector layers already loaded in QGIS, these will be displayed in the next window. You can either choose to process a specific layer by individually clicking on it or, alternatively, you can navigate to where the vector data that needs to be merged resides.

 Note that you can process centroids for points, lines, and polygons. However, we would suggest that processing against a polygon will give you better results than processing against points or lines.

To save the output, click on the browse button at the end of the **Centroids** textbox and choose to save to file (where you can set the vector file type and give it a name). Save to `GeoPackage` or write the data to a `PostGIS` database if you have one set up.

Click on the **Run** button to start the centroid tool. Depending on the size of the task, this could take seconds or minutes to give an output similar to this:

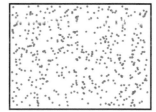

Converting lines into polygons

This is the **Lines to Polygons** window that enables you to convert lines into polygons:

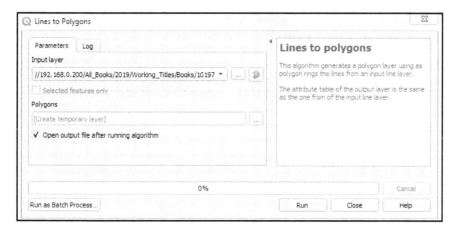

In the **Lines to Polygons** window, set the **Input layer** to be processed by either selecting it from the drop-down list or clicking on the browse button and navigating to the file.

To save the output, click on the browse button at the end of the **Polygons** textbox and choose to save to file (where you can set the vector file type and give it a name). Save to GeoPackage or write the data to a PostGIS database if you have one set up.

Click on the **Run** button to start the **Lines to Polygons** tool. Depending on the size of the task, this could take seconds or minutes. The log tab at the top of the window will show you the processing processes involved in creating the centroid data and indicate whether there are any errors in the task of merging the data.

Converting polygons into lines

You can even convert polygons into lines:

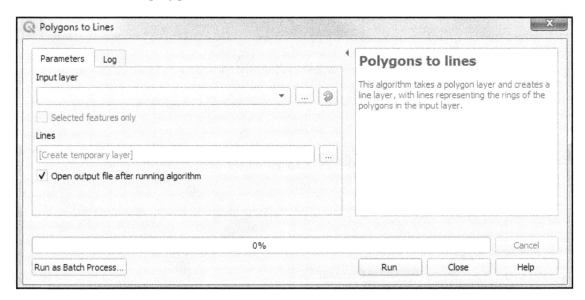

In the **Polygons to Lines** window, set the **Input layer** to be processed by either selecting it from the drop-down list or clicking on the browse button and navigating to the file.

To save the output, click on the browse button at the end of the **Lines** textbox and choose to save to file (where you can set the vector file type and give it a name). Save to `GeoPackage` or write the data to a `PostGIS` database if you have one set up.

Click on the **Run in Background** button to start the **Polygons to Lines** tool. Depending on the size of the task, this could take seconds or minutes. The log tab at the top of the window will show you the processing processes involved in creating the centroid data and indicate whether there are any errors in the task of merging the data.

Creating polygons surrounding individual points

In QGIS, there are two tools for generating polygons around individual points in a layer:

- **Voronoi polygons** (using Voronoi diagram)
- **Delaunay triangulation**

Voronoi polygons

Voronoi polygons represent the area of influence around each point. These are named after the Russian mathematician Georgy Voronoy, who invented the algorithm. They are also referred to as Thiessen polygons and are named after Alfred Thiessen, who independently created the same algorithm.

You can use the Voronoi polygon in QGIS by navigating to **Vector | Geometry Tools | Voronoi Polygons**:

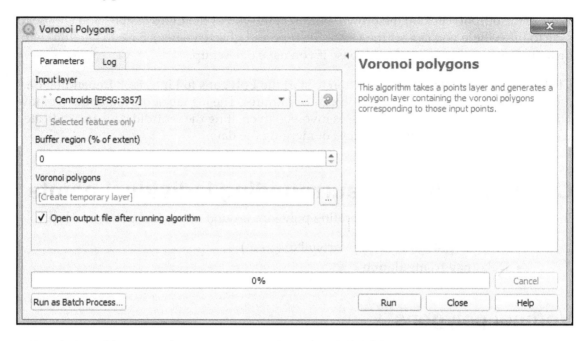

In the **Voronoi Polygons** window, set the **Input layer** to be processed by either selecting it from the drop-down list or clicking on the browse button and navigating to the file.

To save the output, click on the browse button at the end of the **Voronoi polygons** textbox and choose to save to file (where you can set the vector file type and give it a name). Save to GeoPackage or write the data to a PostGIS database if you have one set up.

Click on the **Run in Background** button to start the **Voronoi Polygons** tool. Depending on the size of the task, this could take seconds or minutes. The log tab at the top of the window will show you the processing processes involved in creating the centroid data and indicate whether there are any errors in the task of merging the data.

This is a kind of polygon that can be acquired using Voronoi polygons:

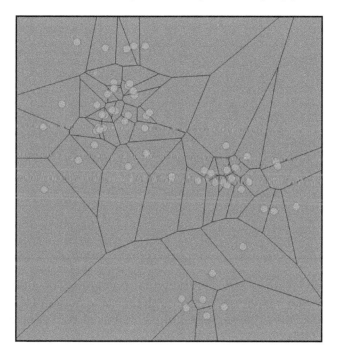

Delaunay triangulation

The Delaunay triangulation tool can be found by navigating to **Vector | Geometry Tools | Delaunay Triangulation**:

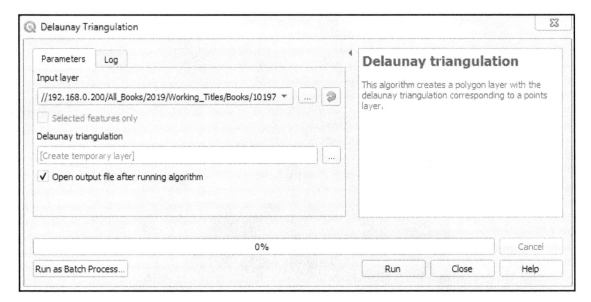

Delaunay triangulation creates a series of triangular polygons. The method creates a triangle in such a way that a circle drawn through the three nodes of the triangle will contain no other nodes. This is the same technique that is used to generate **Triangulated Irregular Networks** (TINs).

In the **Delaunay Triangulation** window, set the **Input layer** to be processed by either selecting it from the drop-down list or clicking on the browse button and navigating to the file.

To save the output, click on the browse button at the end of the **Delaunay triangulation** textbox and choose to save to file (where you can set the vector file type and give it a name). Save to GeoPackage or write the data to a PostGIS database if you have one set up.

Click on the **Run in Background** button to start the **Delaunay Triangulation** tool. Depending on the size of the task, this could take seconds or minutes.

If we set the Voronoi buffer region to 10 meters, this is the type of polygon that we will get:

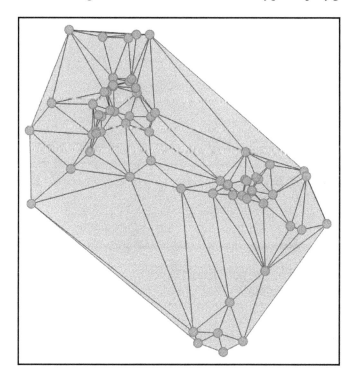

Extracting nodes (vertices)

The extraction of vertices (sometimes referred to as nodes) allows you extract points from lines and polygons. Go to **Vector | Geometry Tools | Extract Vertices**:

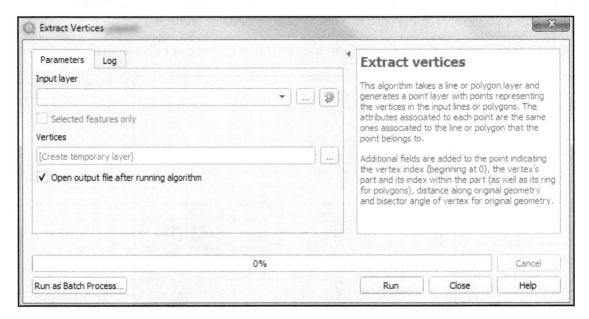

In the **Extract Vertices** window, set the **Input layer** to be processed by either selecting it from the drop-down list or clicking on the browse button and navigating to the file.

To save the output, click on the browse button at the end of the **Vertices** textbox and choose to save to file (where you can set the vector file type and give it a name). Save to GeoPackage or write the data to a PostGIS database if you have one set up.

Click on the **Run in Background** button to start the extraction tool. Depending on the size of the task, this could take seconds or minutes. This will be the output:

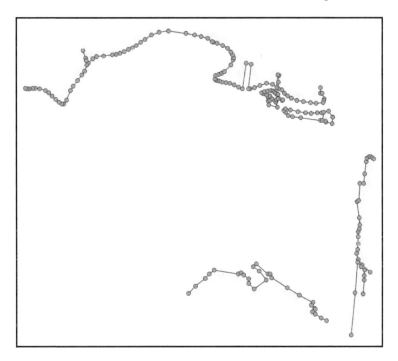

Simplifying and densifying features

The **Simplify geometries** and **Densify geometries** tools (which can found by navigating to **Vector** and **Geometry Tools**) remove (simplify) and add (densify) vertices, respectively. They are only suitable for line and polygon data.

Simplifying data may be desirable to make it more suitable for use at a smaller scale. It may also be helpful if the data is to be used in an online interactive mapping scenario. The simplify tool uses a modified Douglas-Peucker algorithm that reduces the number of vertices while attempting to maintain the shape of the features.

Converting between multipart and singleparts features

In a typical vector layer, one feature corresponds to one record in the attribute table. In a multipart layer, there are multiple features that are tied to one record in the attribute table. To access it, go to **Vector** | **Geometry Tools** | **Multipart to Singleparts**:

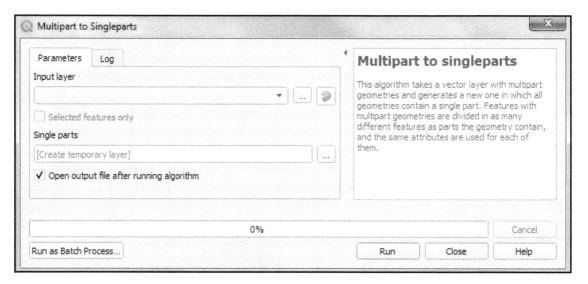

In the **Multipart to singleparts** window, set the **Input layer** to be processed by either selecting it from the drop-down list or clicking on the browse button and navigating to the file.

To save the output, click on the browse button at the end of the **Single parts** textbox and choose to save to file (where you can set the vector file type and give it a name). Save to GeoPackage or write the data to a PostGIS database if you have one set up.

Click on the **Run** button to start the **Multipart to Singleparts** tool. Depending on the size of the task, this could take seconds or minutes.

Adding geometry columns to an attribute table

The geometry tool is an easy to use tool that allows you to add the following geometry values to points, lines, and polygons:

- **Point**: The XCOORD and YCOORD columns will contain the *x* and *y* coordinates of the point
- **Line**: The SHAPE_LEN column will contain the length of the record's line(s)
- **Polygon**: The AREA and PERIMETER columns will contain the area and perimeter of the record's polygon(s)

To access this, go to **Vector** | **Geometry Tools** | **Add Geometry Attributes**.

In the **Add Geometry Attributes** window, set the **Input layer** to be processed by either selecting it from the drop-down list or clicking on the browse button and navigating to the file.

To save the output, click on the browse button at the end of the **Add Geometry Attributes** textbox and choose to save to file (where you can set the vector file type and give it a name). Save to GeoPackage or write the data to a PostGIS database if you have one set up.

Click on the **Run** button to start the **Add Geometry Attributes** tool. Depending on the size of the task, this could take seconds or minutes.

Using basic vector Geoprocessing Tools

This section will focus on **Geoprocessing Tools** that use vector data layers as input to produce derived output.

These tools can be found in the **Geoprocessing Tools** menu under **Vector**. The icons next to each tool in the menu give a good indication of what each tool does.

We will look at some commonly used spatial overlay tools such as clip, buffer, and dissolve. In the case of a simple analysis, these tools may serve to gather all of the information that you need.

In more complex scenarios, they may be part of a larger workflow.

The tools that are covered in this chapter are also available via the **Processing Toolbox**, which is installed by default with QGIS Desktop. When enabled, this plugin turns on the **Processing** menu from which you can open the **Processing Toolbox**. The toolbox is a panel that docks to the right-hand side of QGIS Desktop, and the tools are organized in a hierarchical fashion. The toolbox contains tools from different software components of QGIS such as GRASS, SAGA, and GDAL/OGR, as well as the core QGIS tools that are covered in this chapter.

Spatial overlay tools

Spatially overlaying two data layers is one of the most fundamental types of GIS analysis. It allows you to answer spatial questions and produce information from data. For instance, how many fire stations are located in Portland, Oregon? What is the area covered by parks in a neighborhood?

This series of spatial overlay tools computes the geometric intersection of two or more vector layers to produce different outputs. Some tools identify overlaps between layers and others identify areas of no overlap. The spatial overlay tools include **Clip**, **Difference**, **Intersect**, **Symmetrical Difference**, and **Union**. When using a tool that requires multiple vector input data layers, the layers must be in the same coordinate reference system.

Using the Clip and Difference tools

These two tools are related in that they are the inverse of each other. Data often extends beyond the bounds of your study area. In this situation, you can use the **Clip** tool to limit the data to the extent of your study area. It is often described as a **cookie cutter**. It takes an input vector layer and uses a second layer as the clip layer to produce a new dataset that is clipped to the extent of the clip layer. The **Difference** tool takes the same input, but outputs the input features that do not intersect with the clip layer.

Go to **Vector** | **Geoprocessing Tools** | **Clip**:

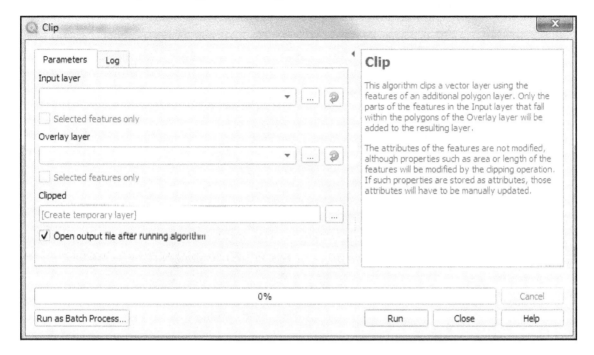

In the **Clip** window, set the **Input layer** to be processed by either selecting it from the drop-down list or clicking on the browse button and navigating to the file.

To save the output, click on the browse button at the end of the **Clipped** textbox and choose to save to file (where you can set the vector file type and give it a name). Save to GeoPackage or write the data to a PostGIS database if you have one set up.

Click on the **Run** button to start the clip tool. Depending on the size of the task, this could take seconds or minutes:

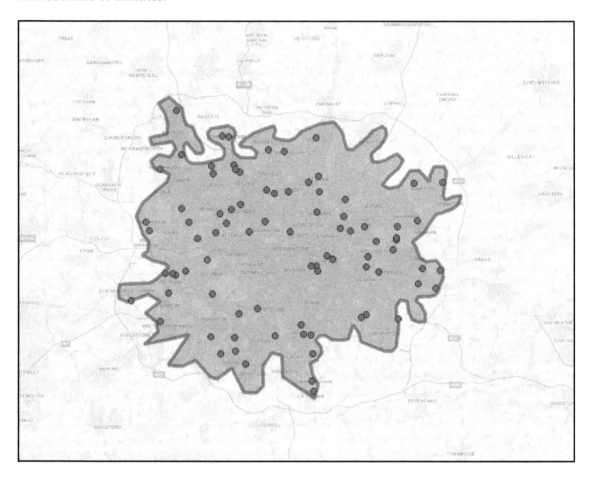

In this example, we have clipped all of the sampling points to the London Urban area vector layer. This data can be found in the samples for this chapter so that you can practice this process.

Go to **Vector** | **Geoprocessing Tools** | **Difference**:

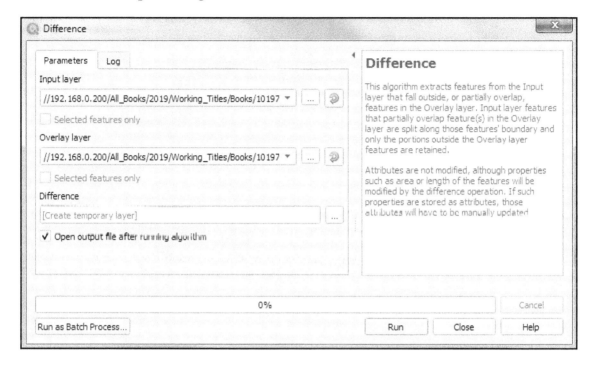

In the **Difference** window, set the **Input layer** to be kept by either selecting it from the drop-down list or clicking on the browse button and navigating to the file. In the **Difference** layer, set the layer that you want to process with the differences against the **Input layer**.

To save the output, click on the browse button at the end of the **Difference** textbox and choose to save to file (where you can set the vector file type and give it a name). Save to `GeoPackage` or write the data to a `PostGIS` database if you have one set up.

Click on the **Run in Background** button to start the **Difference** tool:

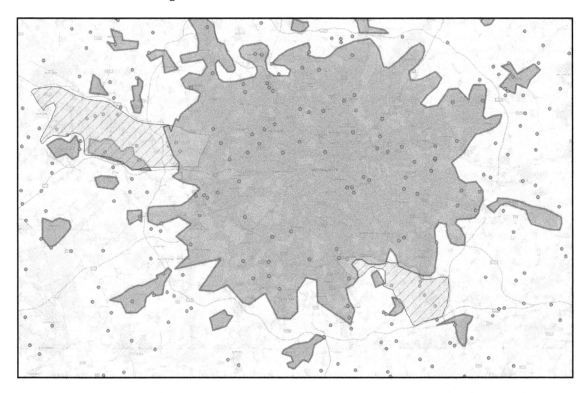

In this example, we are using the London Urban and Special Areas of Conservation vector layers to output a vector layer that shows the differences between the two layers. This data can be found in the GitHub repository associated with this chapter so that you can practice this process.

Using the Intersect and Symmetrical Difference tools

The **Intersect** tool preserves only the areas that are common to both datasets in the output. **Symmetrical Difference** is the opposite; only the areas that do not intersect are preserved in the output.

Unlike **Clip** and **Difference**, the output from these two tools contains attributes from both input layers. The output will have the geometry type of the minimum geometry of the input.

Overlaying polygon layers with Union

The **Union** tool overlays two polygon layers and preserves all of the features of both datasets, regardless of whether or not they intersect.

To access this, go to **Vector** | **Geoprocessing Tools** | **Union**:

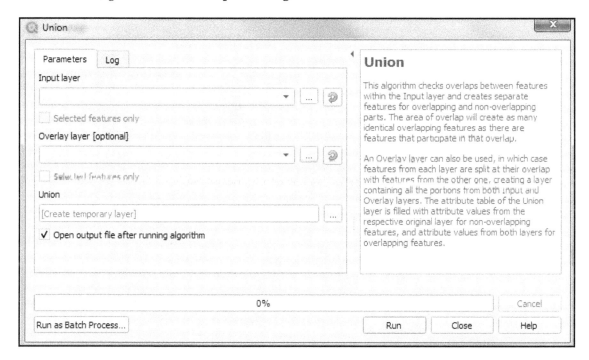

In the **Union** window, set the **Input layer** to be kept by either selecting it from the drop-down list or clicking on the browse button and navigating to the file. In the **Union** layer, set the layer that you want to union with the **Input layer**.

To save the output, click on the browse button at the end of the **Union** textbox and choose to save to file (where you can set the vector file type and give it a name). Save to GeoPackage or write the data to a PostGIS database if you have one set up.

Click on the **Run in Background** button to start the **Union** tool. Depending on the size of the task, this could take seconds or minutes. The following is the result:

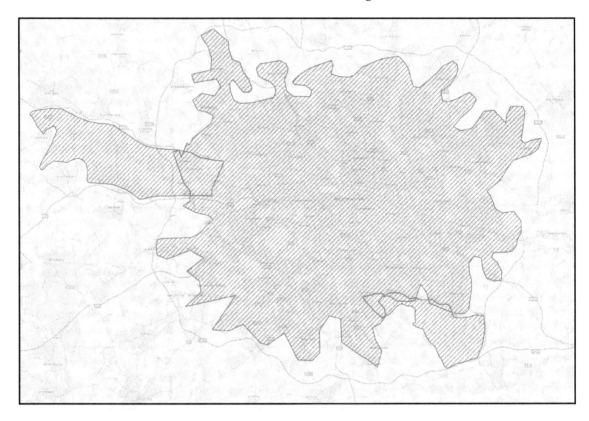

This is an example of the output from a union between the London Urban and Special Areas of Conservation vector layers. These datasets are available in the code repository of this chapter, the link to which can be found in the *Preface*.

Creating buffers

The **Buffer** tool is a commonly used tool that produces a new vector polygon layer which represents a specific distance from the input features. It can be used to identify proximity to a feature.

To access this, go to **Vector** | **Geoprocessing Tools** | **Buffer**:

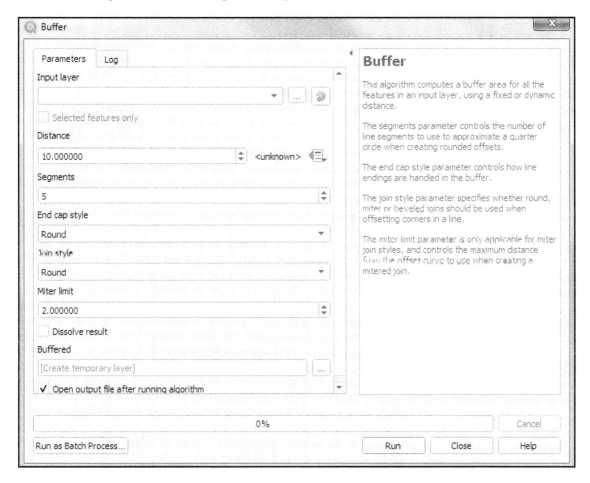

Choose the **Input layer** from the drop-down list or click on the navigation button and load the layer that you want to buffer. Note that you can buffer points, lines, and polygons.

Set the distance—this is set in meters:

- **Segments**: This controls the number of line segments that are used to make the curved buffer. The lower the number, the less smooth the buffer will appear.
- **End cap style**: This sets how lines segment lines are handled.
- **Join style**: This sets the join style parameter and specifies whether round, miter, or beveled joins should be used when offsetting corners in a line.

- **Miter limit**: The miter limit parameter is only applicable for miter join styles, and controls the maximum distance from the offset curve to use when creating a mitered join.
- **Dissolve result**: If ticked, this will merge the buffers that overlap and put them into a single polygon.

To save the output, click on the browse button at the end of the **Buffered** textbox and choose to save to file (where you can set the vector file type and give it a name). Save to `GeoPackage` or write the data to a `PostGIS` database if you have one set up.

Click on the **Run in Background** button to start the **Buffer** tool. This will give the following output:

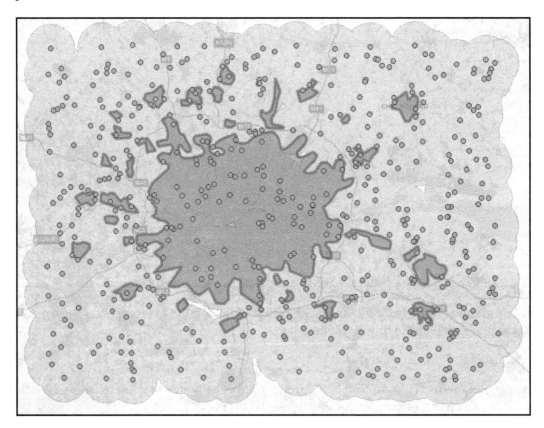

In this example, we have dissolved the buffer output from the sampling points.

Generating convex hulls

The **Convex Hull** tool will take a vector layer (point, line, or polygon) and generate the smallest possible convex bounding polygon around the features. It will generate a single minimum convex hull around the features or allow you to specify an attribute column as input. In the latter case, it will generate convex hulls around features with the same attribute value in the specified field.

To access this, go to **Vector** | **Geoprocessing Tools** | **Convex Hull**. Set the **Input layer** and the **Convex hulls** output file to generate an export, and then click **Run in Background**.

Dissolving features

The **Dissolve** tool merges the features of a GIS layer into one feature. This is useful if you have a large number of features all with the same attribute value.

To access this, go to **Vector** | **Geoprocessing Tools** | **Dissolve.**

Set the **Input layer** and the **Unique ID Field** (which is optional) for the vector file, set the **Dissolved Output** file, and then click **Run in Background**.

Defining coordinate reference systems

QGIS supports hundreds of **Coordinate Reference Systems** (**CRS**) for data display and analysis. In some cases, however, the supported CRS may not suit your exact needs. QGIS provides the functionality to create custom CRS using the **Custom Coordinate Reference System Definition** (**Custom CRS**) tool, which can be found by navigating to **Settings** and then **Custom CRS**. In QGIS, a CRS is defined using the PROJ.4 definition format.

We must understand the PROJ.4 definition format before we can define a new or modify an existing CRS; therefore, in the first part of this section, we will discuss the basics of PROJ.4, and in the second part, we will walk you through an example of how to create a custom CRS. PROJ.4 is another **Open Source Geospatial Foundation** (**OSGeo**) (http://osgeo.org) project that's used by QGIS, similar to OGR, **Geospatial Data Abstraction Library** (**GDAL**), and **Geometry Engine – Open Source** (**GEOS**). This project is for managing coordinate systems and projections. For a detailed user manual for the PROJ.4 that's format used to specify the CRS parameters in QGIS, visit the project website at https://proj4.org/index.html.

Understanding the PROJ.4 definition format

The PROJ.4 definition format is a line composed of a series of parameters separated by spaces. Each parameter has the general form of +parameter=value. The parameter starts with the + character, followed by a unique parameter name. If the parameter requires a value to be set, then an equals sign, =, will follow the parameter name, and the value will follow the equal sign. If a parameter does not require a value to be set, then it is treated as a flag.

The following is a list and discussion of the common parameters that are used when defining a CRS:

- **Projection** (+proj): This is always required. It is the name of the cartographic projection to use. The value that's provided is an abbreviated name of a supported projection.
- **Spheroid** (+ellps): This is the shape of the Earth. Hence, a spheroid is generated using the lengths of the axes, both major and minor, that make the shape look the real Earth. Hence, these models are the best to show many locations on Earth.
- **Datum** (+datum): This is the name of the spheroid to use.
- **Central meridian** (+lon_0): The center/origin(x) of the map lies on this longitude.
- **Latitude of origin** (+lat_0): This is the latitude on which a map is centered (y origin).
- **False easting** (+x_0): This is the x coordinate value for the central meridian (x origin).
- **False northing** (+y_0): This is the y coordinate value for the latitude of origin (y origin).
- **Standard parallel(s)** (+lat_1, +lat_2): The center/origin(y) of the map lies on this latitude.
- **No defaults** (+no_defs): This is a flag to designate that no default values should be utilized for parameters that are not specified in the projection definition.
- **Coordinate units** (+units): These are used to define distances when setting x and y coordinates. For a full list of parameters, visit the PROJ.4 project website at https://proj4.org/index.html.

Defining a new custom coordinate reference system

There are two methods for creating a custom CRS: write a PROJ.4 definition from scratch or copy the PROJ.4 definition from an existing CRS and modify it. No matter which creation method you choose, both are completed using the Custom CRS.

Setting Definitions

To access the **Definition** window, go to **Settings** ⏐ **Custom Projections:**

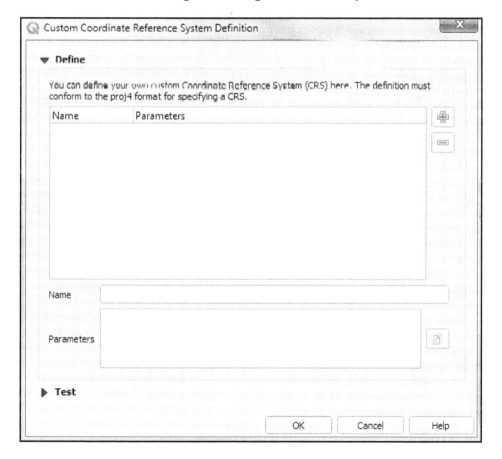

This window has two parts: **Define** and **Test**. We will not use or discuss the **Test** part; instead, we will focus on the **Define** part of the window to create our new CRS. We will modify the USA_Contiguous_Albers_Equal_Area_Conic EPSG:102003 projection so that it focuses on New England.

Click on the **Add new CRS** button to create a blank CRS entry. Set the name of the new CRS to New England Albers Equal Area Conic.

At this point, we have two options: we can write the PROJ.4 projection from scratch in the **Parameters** textbox, or we can copy an existing CRS PROJ.4 string from a projection that closely matches what we want and then modify it to our needs. Let's elect to copy an existing CRS and modify it.

Click on the **Copy existing CRS** button, which will open the **Coordinate Reference System Selector** window. Enter 102003 in the **Filter** textbox to find the USA_Contiguous_Albers_Equal_Area_Conic projection. Select the found projection and then click on **OK**. This will copy the PROJ.4 string back to the **Parameters** textbox in the **Custom Coordinate Reference System Definition** window. In the **Parameters** textbox, modify the PROJ.4 string by changing it to +proj=aea +lat_1=42.5 +lat_2=45 +lat_0=43.75 +lon_0=-71 +x_0=0 +y_0=0 +datum=NAD83 +units=m +no_defs.

Click on **OK** to close the window and store your new custom CRS.

With the creation of the custom CRS, we can apply it as our project CRS to perform an on-the-fly CRS transformation (by navigating to **Project |Project Properties | CRS**).

The new custom CRS can be found at the bottom of the CRS.

Viewing a statistical summary of vector layers

It is often useful to have an at-a-glance view of a vector layer's statistics when you are working on a project. QGIS has a **Statistics Panel** that allows you to view a long list of descriptive statistics for whichever vector layer you choose. In this section, we will review the capabilities of the statistical summary tool in QGIS.

To open the **Statistical Summary** tool, either click **View | Statistical Summary** or **View | Panels | Statistics Panel**.

This will open the **Statistics Panel**. Like all other panels, it can float on the screen or be set to dock, so you may need to search your screen to find the panel.

Advanced field calculations

QGIS Desktop provides powerful field calculation functionality. In the field calculator, advanced mathematical, geometry, string, date and time, type conversion, and conditional functions are available for use. Leveraging these advanced functions along with standard operators allows for some powerful field calculations.

This section will explain the field calculator interface in detail, followed by multiple examples of advanced field calculations from a variety of functional areas. It is assumed that you know the basics of field calculations and common operators.

Exploring the field calculator interface

The field calculator can be opened in three ways, which are as follows:

- Open the attribute table of the layer whose details you wish to calculate, and then

  click on the **Open Field Calculator** button ( ) on the attribute table toolbar
- Open the attribute table of the layer whose details you wish to calculate and then press *Ctrl + I* on your keyboard
- Select the layer whose details you wish to calculate in the **Layers** panel and then click on the **Field Calculator button ()** on the attributes toolbar

The **Field calculator** window has five sections:

- **Field Designation**: This determines which field will hold the output of the expression. You can use **Create a new field** or **Update existing field** by selecting the desired option and setting the relevant option(s). A virtual field can also be created by selecting **Create a new field** and **Create virtual field**. A virtual field is not stored in the dataset; instead, it is stored as an expression in the QGIS project file and will be recalculated every time the field is used.
- **List function**: This contains a tree of field calculation functions available for insertion into the expression.
- **Help function**: This displays the help documentation for the selected function in the function list.

- **Operators**: This ensures quick button access to insert commonly used operators into the expression. These operators are also in the function list under the **Operators** branch.
- **Expression**: This is an editable text area that contains the expression which will calculate field values. Underneath the expression is a preview of the output for a sample record. If the expression is invalid, a notice will appear with a link to more information about the expression error

The expression must meet strict syntax guidelines, otherwise the field calculator will report a syntax error instead of an output preview. The following are common syntax rules for expressions:

- Operators should be placed without any special formatting. For example, `+`. Fields should be surrounded by double quotes, for example, `"State_Name"`.
- Text (string) values should be surrounded by single quotes, for example, `'Washington'`.
- Whole numbers (integer) and decimal numbers (float) should be entered without any surrounding characters, for example, `153.27`.

Functions come in two types, as follows:

- **Functions requiring parameters**: These begin with a function name, followed by a set of parentheses. Inside the parentheses are function parameters separated by commas, for example, `log (base, value)`.
- **Functions not requiring parameters**: These begin with a dollar sign (`$`) followed by the function name, for example, `$area`.

If this is a little confusing, don't worry: you can rely on the field calculator to enter a portion of the syntax for you correctly. To add an operator, field, or function to the expression, double-click on the desired item in the function list and it will be added to the cursor location in the expression.

In addition to adding functions through the function list, the field calculator can also add any value that currently exists in any field to the expression. To do this, expand the **Fields and Values** branch of the function list tree. A list of the fields in the attribute table will be listed.

Writing advanced field calculations

Let's put what we learned previously to practice. This section will walk you through creating three advanced field calculations. The first calculation will insert the current date into a field as a formatted string. The second calculation will insert a geometry value. The third calculation will calculate a label string that differs depending on the state's population.

Calculating and formatting current date

The first example of an advanced field calculation uses two functions to calculate and format the current date. For this example, we will format the current date as dd/mm/yyyy:

1. Load `sampling_points.shp` into the map canvas, which can be found in your demo data.

2. Open the Field calculator icon (⬜)

3. Select **Create a new field** and set the following options:

   - **Output field name**: `Updated`
   - **Output field type**: `Text (string)`
   - **Output field width**: `10`

4. In the **Function list** field, expand the **String** node and then double-click on **format_date** to add it to the **Expression** area. This function takes two arguments: a time string and a string representing the format to convert the time string into. We will use the current date function for the time and write a format string. In the **Function list** field, expand the **Date and Time** node and then double-click on **now** to add it to the **Expression** area after the open parenthesis. Type a comma after **now()** and enter `'dd.MM.yyyy'`, followed by a closed parenthesis. It should look like this : `format_date( now(),'dd.MM.yyyy')`. The `now()` function returns a string representation of the current time and date.

5. Click on **OK** to execute the calculation. This will enable editing on the layer and calculate the field values. Open the attribute table of the layer to see the calculated and formatted date.

6. To make the calculation permanent, save the edits to the layer and disable editing mode.

Calculating with geometry

The second example of an advanced field calculation uses two functions to insert the centroid x coordinate of a geometry object and the number of vertices that compose a geometry object. First, we will calculate the x coordinate of the centroid. To do this, perform the following steps:

1. Open the Field calculator icon (🧮).
2. Select **Create a new field** and set the following options:

   - **Output field name**: XCoord
   - **Output field type**: Decimal number (real)
   - **Output field width**: 10
   - **Output field precision**: 7

3. In the **Function list** field, expand **Geometry** and then double-click on **x** to add it to the **Expression** area. Next, double-click on **$geometry** to add it after the open parenthesis. $geometry represents the geometry of the row being calculated. Lastly, close the parenthesis by typing) on your keyboard. The completed expression should be x($geometry). This function returns the x coordinate of a point geometry or the centroid X coordinate of a non-point geometry (for example, line or polygon).
4. Now, let's store the number of vertices that compose each geometry object in the layer. Open the Field calculator icon (🧮)

   Select **Create a new field** and set the following options:

   - **Output field name**: NumVerticies
   - **Output field type**: Whole number (integer)
   - **Output field width**: 4

5. In the **Function list** field, expand **Geometry** and then double-click on **num_points** to add it to the **Expression** area. This function returns the number of vertices that compose the geometry object. Inside the parentheses, you will need to specify the geometry of which you wish to count the number of vertices. Double-click on **$geometry** to add it to the expression. Close the parenthesis by typing) on your keyboard. The completed expression should read num_points($geometry).

Operators

There are a number of common operators that are good to quickly use to analyze data and produce a result of 1 in the column, with 1 meaning true.

Examples of this might be that you want to update a column with a true value (1) if the area of the polygon is more than 11,000,000 meters squared, `$area > 11000000`, if the number of node points in the geometry is over 10, `num_points( $geometry ) > 10`, or if the perimeter of the polygon is less that 50,000, `$perimeter < 50000`.

You can move beyond this with conditions.

Conditions

Load `urban_other.shp` into the map canvas. Let's create a new virtual field called `area` and calculate the area for each of the polygons. To do this, follow these steps:

1. Click on the Field calculator icon ( ).
2. Select **Create a virtual field** and set the following options:

    - **Output field name**: `area`
    - **Output field type**: `Whole number (integer)`
    - **Output field width**: `10`

3. Use the **$area** function in the function window and then click on **OK**. This will calculate the area in meters squared. If you inspect the attribute table now, you will see a new column with the area field calculated. It is important to look at these value for the next field calculation that we are going to do. This uses a conditional **CASE** function.

Conditionals

Click on the **Field Calculator** for `urban_other.shp` and then select **Create a virtual field**

Set the following options:

- **Output field name**: `size`
- **Output field type**: `Text(string)`
- **Output field width**: `10`

Add the following to the function field:

```
CASE
WHEN $area < 5000000 THEN 1
WHEN $area > 5000000 AND $area < 11000000 THEN 2
WHEN $area > 11000000 THEN 3
ELSE 'investigate'
END
```

What is this doing? This is applying some rules or conditions against the data, which in this instance is the area or **$area** of the polygon. It reads like a sentence: when the area is less than 5,000,000' then input 1 into the column; when the area is greater than 5,000,000 and less than 11,000,000, then add 2 into the column; when the area is greater than 11,000,000, then add 3 into the column; and finally if none of these apply, then the answer is going to be populated with `investigate`.

Summary

Data is rarely in the form that it's needed in to perform processing and analysis. Often, data needs to be merged, checked for validity, converted, calculated, projected, and so on, to make it ready for use. This chapter covered many common preparation tasks to convert raw data into a more usable form.

In the next chapter, the theme of data preparation will continue, but it will be applied to raster data. You will learn how to mosaic, reclassify, and convert raster data to make it more meaningful as input for processing tasks.

Preparing Raster Data for Processing

5

In this chapter, we will be looking at some simple raster tools that enable you to quickly process your data. These processes are not definitive, as there are lots of raster types (aerial, satellite, elevation, and georeferenced images), and it is beyond the scope of this chapter to go into detail about each of these types of rasters.

You will also see that there is no one-rule-fits all solution in the processing tools that we cover. To that end, we suggest that you experiment and play with (where necessary) writing the raster output as temporary layers to the map canvas and then export the result that gives you the best fit for your data requirements.

In this chapter, we will have hands-on exercises on the following topics:

- Merging rasters
- About converting raster files
- Clipping a raster
- Converting a raster into vector
- Converting from vector to raster
- Reclassifying rasters

Merging rasters

In most cases, rasters are supplied in tiles; if there are lot of them, this means that you need to think about how you are going to manage them. Sometimes, the best practical measure to do this is to merge the rasters together into a single file. This can present some complications, normally in terms of the size of the merged raster output.

In a new QGIS project, load in the raster layers SU55SW, SU55SE, SU55NW, and SU55NE from the repository for this chapter, the link of which is in the Preface. Then, follow these steps:

1. Go to **Raster** | **Miscellaneous** | **Merge**:

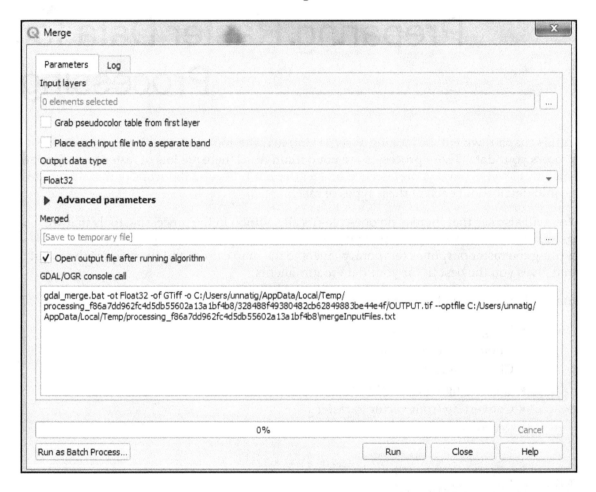

2. In the Merge window, choose the files that you want to merge from the Input layers browser button (SU55SW, SU55SE, SU55NW, and SU55NE).

3. Check the **Grab pseudocolor table from first layer** option. This will make sure that the right color bands are applied across all of the merged files. Keep all other values and settings as listed.

4. Set the **Merged** field to be the location, name, and file type for the new merge raster, or leave it blank to write out a temporary layer called **Merged**.

   There is a **GDAL/OGR console call** text box at the bottom of this window that allows you to see the GDAL commands that will be used. As you become more familiar with GDAL and the use of command-line scripting, you can copy and paste this GDAL command and run it outside of QGIS.

5. Once the file has been merged, it will be displayed on the map canvas. If the raster is presented in black and white, then the easiest method to apply the correct rendering is to copy the style from one of the original raster tiles and paste it against the merge file.

6. Right-click on the original tile, and go to **Style** | **Copy Style**. Now, right-click on the merge file and go to **Style** | **Paste Style**.

It is not always the case that when using the raster tools (as outlined in this chapter) that you have to have the raster layers loaded onto the map canvas. By using the Browser button, which is located at the end of the Layer Input file, you can select a file to be worked on that hasn't been loaded onto the map canvas.

About converting raster files

Sometimes, it is necessary to convert raster file types. The process that's required to convert the data may need to be run through a number of times so that you can produce the best quality image. There are two options to do this: using the Translate tool or Exporting Raster.

Translating

In the following example, we have taken an OpenData aerial photograph from the UK Environment Agency, which is located in the aerial folder in the samples. It has been processed twice to demonstrate why it might be necessary to experiment with settings. As you can see in the following screenshot, converting the data from ECW to TIFF has degraded the image and, in both instances, increased the file size of the image:

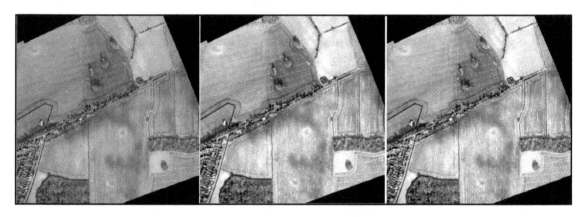

| Original ECW (13mb) | Default conversion to TIF (381mb) | High Definition conversion to TIF(77mb) |

To convert the data through this method, follow these steps:

1. Go to **Raster | Conversion | Translate (Convert Format)...**:

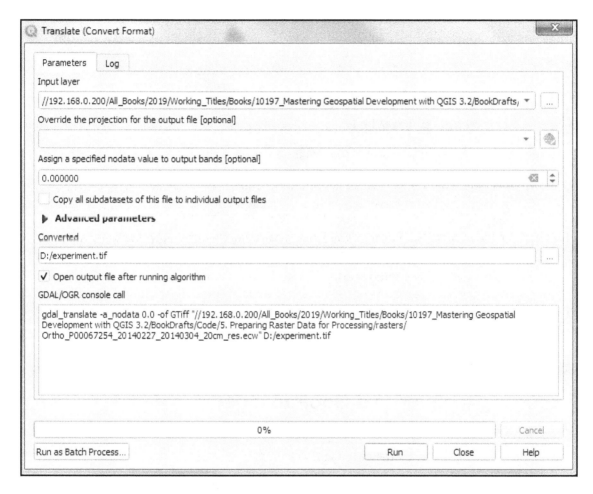

2. In the **Translate (Convert Format)** window, choose the files that you want to convert via the Input layers browser button.
3. Change the projection (if required) by choosing the Override the projection for the output file [optional] option.

4. If the data comes with metadata specifying what the no data value is, then add this in. No data values on a raster are typically solid black areas, as seen in the examples in the Raster conversion section. In the preceding instance, the No Data value is -9999.
5. Change the profile to No Compression (but you might want to experiment).
6. Set the Converted field to be the location, name, and file type for the newly converted raster or leave it blank to write out a temporary layer.

Exporting to a raster

An alternative to the Translate tool is to save the image via the Export options via the layer:

1. Right-click on the raster layer and go to **Export | Save As…**.
2. In the **Save Raster Layer as...** window, change the Output mode to Rendered image.
3. Change the Format to the file type you want and then set a **File Name** and **Location** of the new raster file type by clicking on the browse button.
4. Under **Extent (current: layer)**, click on the drop-down option on the Calculate from Layer button and choose the layer you originally wanted to export into a new raster type.
5. Leave all other options as listed.
6. Click **OK**.

Exporting a raster to a GeoPackage

This might be an option if you want to share your data as a more transportable format.

Follow these steps to export a raster to GeoPackage:

1. To save a raster, right-click on the raster layer and go to **Export | Save As…**.

   A GeoPackage can only have a single layer within it. Merge your series of rasters before saving them to a GeoPackage.

2. Change the **Output mode** to **Rendered image**.
3. Change the **Format** to **GeoPackage** and then set a **File Name** and **Location** for the new GeoPackage by clicking on the browse button.
4. Under **Extent (current: layer)**, click on the drop-down option on the **Calculate from Layer** button and choose the layer you originally wanted to export into a new raster type.
5. Leave all other options as listed.
6. Click **OK**.

The process to write the raster to the GeoPackage can take some time.

Clipping a raster

It may be a requirement to only work on a small or specific area of a raster image. To achieve this, we can clip a raster in a few ways:

1. Go to **Raster | Extraction | Clip Raster by Extent**.
2. Set the **Input Layer** via the drop-down or the browser button.
3. Set the **Clipping Extent**. This is best achieved via the options button at the end of the field and choosing **Select extent** on canvas. You'll then be able to draw a rectangle on the map canvas, defining the extent that you want to click. This will be translated into numerical values in the clipping extent text box.
4. Set the **Clipped (extent)** field on the location, name, and file type to be stored. You can leave this blank, which means that a temporary file will be written to the map canvas.

When you go to **Raster | Extraction | Clip Raster by Mask Layer...**, you'll get the following output:

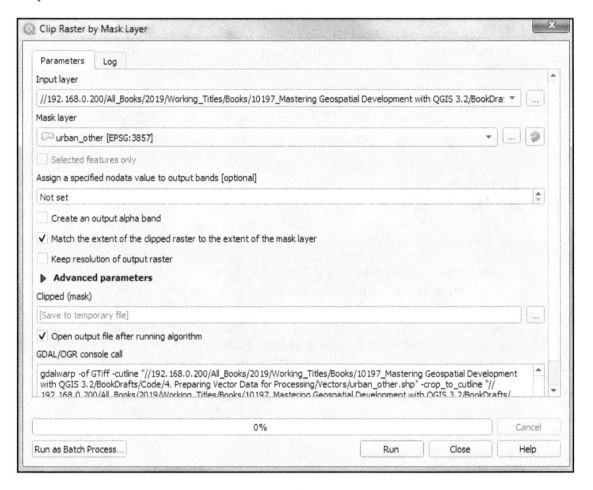

If you have a vector layer loaded onto the map canvas or stored in a file directory, you can use this to define the extent of the area you want clipped. If you don't have such a vector file, then you can quickly create a temporary scratch layer by clicking on Layer | Create Layer | New Temporary Scratch Layer. You can then use this layer to capture the extent that you want to clip the raster to.

In a new QGIS project, load in the following layers from the sample data:

- **Ortho_P00067266_20140227_20140304_20cm_res.ecw**
- Extract by **clip.shp**

To acquire a raster output, perform the following steps:

1. Go to **Raster | Extraction | Clip Raster by Mask Layer**.
2. Set the Input Layer via the drop-down menu or the browser button (**Ortho_P00067266_20140227_20140304_20cm_res**).
3. Set the **Mask Layer** (Extract by clip).
4. Set the **Clipped (mask)** field on the location, name, and file type to be stored. You can leave this blank, which means that a temporary file will be written to the map canvas.
5. Click on **Run in Background**.

This will give you a clipped raster, as shown here:

Converting rasters into vectors

Rasters can have limitations when it comes to querying them, and so it might be necessary to export a raster to a vector layer from time to time.

While this tool is very useful, we would urge you not to convert rasters that cover a large geographic area. It may be better to process individual tiles into vectors and then merge the vectors together at a later time.

Using the output from the previous section (if this is a temporary file, this will be called Clipped (Mask)), perform the following steps:

1. Go to **Raster | Conversion | Polygonize (Raster to vector)**.
2. In the Polygonize (Raster to Vector) window, choose the Input Layer via the drop-down menu or Browser button. Choose **Clipped (Mask).**
3. Choose the **Band number (Band 1 (Gray))**. You might want to experiment to see what the other bands produce.
4. You can change the name of the field to be created, which will contain the Band value if you wish. The default is **DN**.
5. Set the vectorized file location, name, and file type if required; otherwise, leave it blank and write the file out as a Temporary file.
6. Click **Run in Background**.

Once you have the new vector layer in the map canvas, you can change the styling of the layer; we would suggest a categorized styling based on DN (the default band value field).

This is an example of the output to a vector from a raster:

Converting from vector to raster (rasterize)

To convert a vector to a raster format, QGIS provides the Rasterize tool. This tool converts a shapefile to a raster and applies the values in a specified attribute field to the cell values. To access the Rasterize tool, click on **Rasterize (Vector to Raster)** by navigating to **Raster | Conversion**.

The Rasterize tool, shown in the following screenshot, uses the **Dallas_Land_Cover.shp** sample file as input:

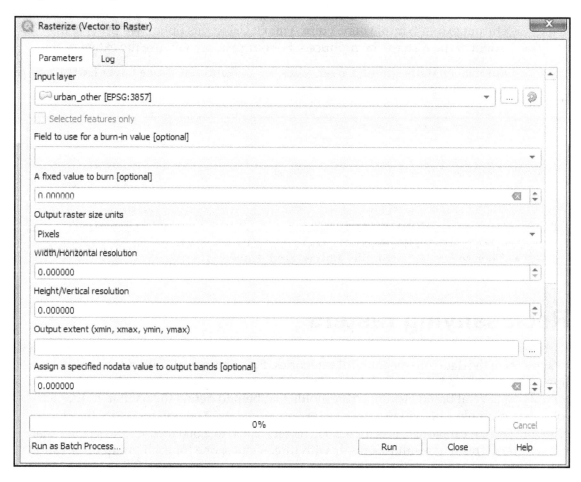

To convert a raster to a vector polygon, some of the available options available are as follows:

- **Input layer**: The input vector file to be converted. The tool supports multiple vector formats.
- **Field to use for a burn-in value**: The field name from the polygon to be burned into raster.
- **Output raster size units**: This tells the unit of raster size.

- **Width/Horizontal resolution**: This is the width of the raster file or the output file.
- **Height/Vertical resolution**: This is the height of the raster file or the output file.
- **Output extent (xmin, xmax, ymin, ymax)**: Using this option we provide the extent of the raster to be produced. For our case, we will use the extent of the input layer **Dallas_land_Cover**. Click on 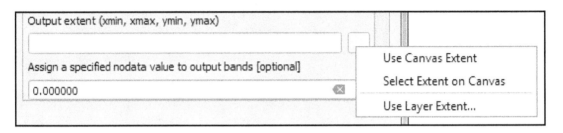 and select **Use Layer Extent...** as shown in the snapshot:

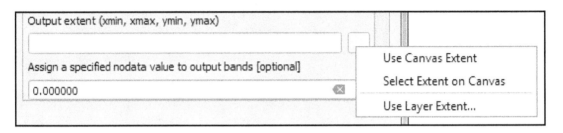

- Then select the vector file, **Dallas_land_Cover.**

Reclassifying rasters

Raster datasets often have hundreds or thousands of values. For an analysis, you may need to synthesize the data into meaningful categories. For example, elevation may be an important input in a habitat model for species X. However, you may only be interested in identifying several broad elevation thresholds that help to define the habitat. In the following example, you will use the `elevation.tif` data. You will reclassify the elevation data into several categories: less than 2,000 meters, 2,000 to 2,500 meters, and greater than 2,500 meters. This will result in a raster with three values, one for each group of elevation values.

The following steps outline how to use the **r.recode** GRASS tool (found in the Processing Toolbox) to accomplish this.

If r.recode doesn't open after double clicking or shows error, then close **QGIS Desktop 3.6.0 (or QGIS Desktop 3.4)** and open and use **QGIS Desktop 3.6.0 with GRASS 7.6.0 (or QGIS Desktop 3.4.5 with GRASS 7.6.0)** to use this **GRASS** tool .

1. Load **elevation.tif** and set the project's **CRS** to **EPSG: 26912**.
2. Turn on the processing plugin (by navigating to **Plugins | Manage and Install Plugins**) if it is not enabled.
3. Open the Processing panel by clicking on **Toolbox** under **Processing**. The **Processing Toolbox** is covered in more detail in Chapter 8, *The Processing Toolbox*.
4. To help locate the tool, type recode into the Processing Toolbox search bar and hit the Enter key. Double-click on the tool to open it.
5. Select the input layer by clicking on the down arrow to choose a raster loaded in the canvas or by clicking on the browse button.
6. Next, the tool will ask for a value to be filled in the File containing recode rules [optional] field. This file has to be created in a text editor. The syntax for the recode rules file is as follows:

```
input_value_low:input_value_high:output_value_low:output_value_high
input_value_low:input_value_high:output_value
*:input_value:output_valueinput_value:*:output_value
```

7. The following are the recode rules for this example. The first line tells the tool to recode the values less than 2000 meters with a value of 1 in the output raster. The first asterisk is a wildcard for every value less than 2000. he second line recodes the values greater than and equal to 2000 and less than 2500 as 2 in the output raster. The third line recodes all values greater than 2500 as 3 in the output raster:

```
*:2000:1
2000:2500:2
2500:*:3
```

8. Save the preceding code to a text file named `Elevation_rRecode_Rules.txt`.
9. Select the output raster by clicking on the browse button. You can choose either to **Save to a temporary file** or **Save to file**. The following screenshot shows the completed **r.recode** tool:

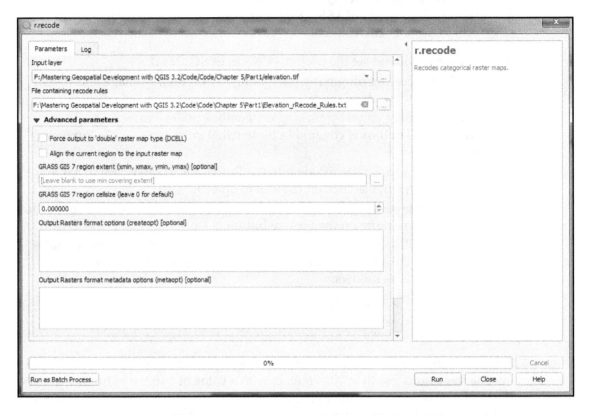

The following figure shows the result of the reclassification. The original elevation raster with the original elevation values is on the left, and the reclassified raster with three values is on the right:

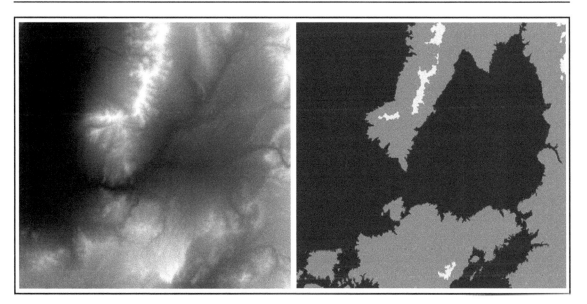

There is a similar GRASS tool in the Processing Toolbox named r.reclass. The r.reclass tool is used when reclassifying integer and categorical rasters, while r.recode will reclassify floating-point and decimal value rasters. Both tools use the same format for the rules text file. More complete documentation for these tools can be found on the GRASS GIS help pages at `https://grass.osgeo.org/grass77/manuals/r.reclass.html` and `https://grass.osgeo.org/grass77/manuals/r.recode.html`.

Summary

In this chapter, we have looked at preparing rasters for further analysis/processing work. We have also seen that, while demonstrating the tools and processes that can be applied to raster images, the settings that are required to produce the end result that you wish are beyond the scope of this chapter, as there are too many variables, file extension types, and indeed raster image types to cover everyone's needs. Working with temporary files gives you a lot of flexibility to test different processes and settings.

In the next chapter, we will see what new features QGIS has for editing and ways to create different types of data. We will also work with some geocodes and mappings.

3
Section 3: Diving Deeper

In this section, we will do advanced operations on QGIS data. We will see how we can use QGIS tools to create, edit, and visualize complex data.

We will cover the following chapters in this section:

- Chapter 6, *Advanced Data Creation and Editing*
- Chapter 7, *Advanced Data Visualization*

6
Advanced Data Creation and Editing

This chapter will provide you with a number of advanced ways to create vector and raster data. We will see explanations and step-by-step examples of mapping raw coordinate data, geocoding address-based data, georeferencing imagery, validating vector data with topological rules, and topological editing. With the topics covered up to this point, you will be able to work with a variety of vector, raster, and tabular input data. You will also see advanced ways to create vector and raster data. There is a great deal of spatial data held in tabular format. Readers will learn how to map coordinate-based and address-based data. Other common sources of geospatial data are historic aerial photographs and maps, in a hardcopy format. You will learn how to georefer scanned imagery and transform it into a target coordinate reference system. The final part of this chapter will cover testing topological relationships in vector data and correcting any errors via topological editing.

This chapter will cover the following topics:

- What's new in editing
- Creating points from coordinate data
- Geocoding address-based data
- Georeferencing imagery
- Checking the topology of vector data
- Repairing topological errors via topological editing
- Repairing an overlap between polygons
- Repairing a gap between polygons

What's new in editing?

In QGIS 3.2, QGIS 3.4, and QGIS 3.6, new editing features have been added. These features make editing in QGIS much easier, and they allow users to perform more sophisticated and user-controlled data creation and editing.

CAD-style digitizing tools

Let's have a look at the new CAD-style digitizing tools by looking at different options under **Edit** in the **Menu** bar. Here, we will create and edit both a **New Shapefile Layer** and a new **Temporary Scratch Layer** (in-memory layers are not saved and get removed when we close QGIS).

Next, we will discuss features that are completely new to QGIS 3.6, QGIS 3.4, and QGIS 3.2.

Adding a circle

We can easily create and edit a circle using the **Add Circle** option, which is under the **Edit** option. It has many variants for creating a circle, as you can see from the following screenshot:

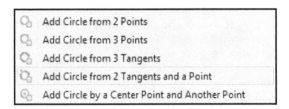

We will now create a new shapefile and then create polygon data. We create a shapefile by clicking on **Layer** | **Create Layer** | **New Shapefile Layer...** Now, follow these steps as mentioned to add a new shapefile layer:

1. We give it a new name by browsing to a folder and giving it a name
2. Select **Polygon** for the geometry type
3. Select **EPSG:4326 - WGS 84**
4. If we want to add a new field, write the name of that field; here, we name it value

5. This field has a **Type** of **Decimal Number**
6. Specify the length of this field; here, we write 20
7. Now, click on **Add to Fields List**
8. Click **OK**:

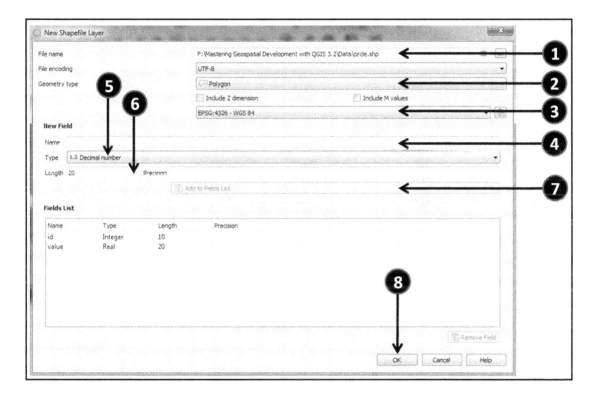

Let's now look at some of these new features.

Adding a circle from two points

Click on **Toggle Editing** ( ) under **Menu Bar** to start editing. Now, click on **Add Circle from 2 Points** under **Edit | Add Circle**. Now, as soon as we move the cursor to the **Map Window** area, it changes its shape to a compass. Click on any point in the **Map Window** area, and now, as we move the cursor in a different direction, a circle grows or shrinks accordingly. When we are satisfied with the shape, right-click to fill the field values.

Now, the **Feature Attributes** window will appear. Write the appropriate value for the **id** and **value** fields and click **OK**:

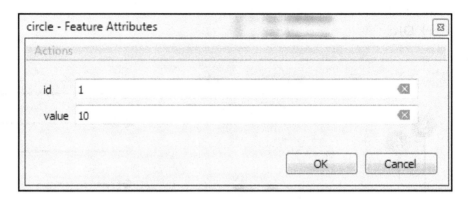

Now, we find a circle, drawn as follows:

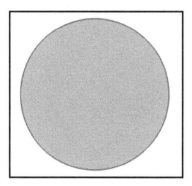

Adding a circle from three points

Here, you first click on the **Map** window to get the first point; then, you can move freely to any other point in a straight-line manner. If you want to use any point in the **Map** window as the second point, you need to click again, and now you will see that a circle moves around according to the cursor. When we are satisfied with the shape, just right-click, and again this will give us the **Feature Attributes** window, which we will use to create a new circle.

Adding a rectangle

Now, you will learn how to create rectangles using the **Add Rectangle** feature of QGIS 3.2 and QGIS 3.4. It provides three options, which are shown as follows:

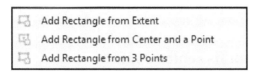

Here, you will be using the **temporary scratch layer** option, but you can also use **Shapefile**, **Geopackage Layer**, and **SpatiLite Layer**. You can create a temporary scratch layer by clicking on **New temporary scratch layer** (   ) or by clicking on **New temporary scratch layer...** under **Layer** | **Create Layer** in the menu bar. Now, you will get a new window where you can name the layer (here, **Rectangle**, but you could name it anything), select **Polygon** as the **Geometry type**, and then click **OK**:

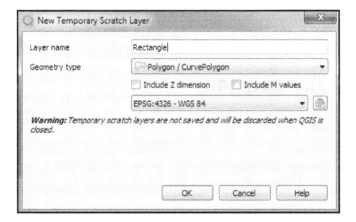

Now, for convenience, make the rectangle layer as the active layer by deselecting any other layers, as shown:

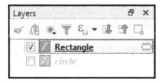

Now, we are ready to create and edit the vector data.

Adding a rectangle from Extent

Now, click on **Add Rectangle from Extent** under **Edit** | **Add Rectangle**. Click on any point in the map window. Starting from this point, you can change the size of a rectangle. Now, as you move the cursor around, the rectangle moves around accordingly, by either growing or shrinking. When you are satisfied with the shape of the rectangle, right-click to finalize the rectangle.

Adding a rectangle from its center point

Here, the first point clicked works as the center of the rectangle, and, as we move around, the rectangle grows or shrinks by making the first-clicked point the center of the rectangle. Now, when we are satisfied with the rectangle's shape and size, just right-click to finish.

Adding a rectangle from three points

Here, the first two clicked points define one side of a rectangle, and as we move around the cursor, the area of the rectangle grows or shrinks by keeping the first side of the rectangle intact and by also having the opposite side having the same length. As before, by right-clicking, we finalize the shape and the size of a rectangle.

Adding a regular polygon

Create a new temporary scratch layer and create a polygon in this layer. Under **Edit** and under **Regular Polygon**, there are three options, as shown here:

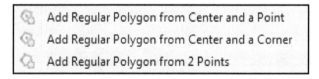

You will now work with the last two options.

Adding a regular polygon from the center and from a corner

The first point we click works as the center of the polygon and, as we move the cursor, this cursor point works as one of the corners of the polygon. At the same time, you will also find an option for changing the number of sides for this polygon at the top-right section of the **Map** window. As soon as we are satisfied with the shape and the size, right-click to create it:

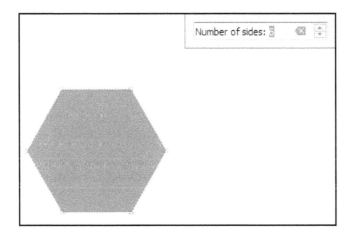

Adding a regular polygon from two points

The first point clicked works as one corner of a side of a polygon, and as we move the cursor, the line between the first point clicked and the cursor position works as one side of the polygon. As we move the cursor, this side of the polygon moves and so does the polygon's size. We also can determine the number of sides here by specifying them as we did in the previous example. Again, right-click to finalize the shape.

Vertex tool

A QGIS version older than version 3 has a node tool ( ) that is now being replaced with a vertex tool ( ), as found in QGIS 3.2 and QGIS 3.4. Using this tool, you can select single or multiple points or vertices at once and move, add, or delete them. This tool is under the menu bar, and we click on the vertex tool ( ) to activate it. Again, we first create a temporary scratch layer and then start editing a polygon.

Let's suppose you have the following polygon and want to add a new point or vertex:

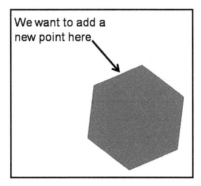

We do so by clicking on the vertex tool and then double-clicking on the location where we want to add a new point. This will add a new point to that place.

Now, we can also move any point by clicking and releasing it and then moving it to where we want. As you see, the newly added point has been moved using this vertex tool by clicking and dragging it:

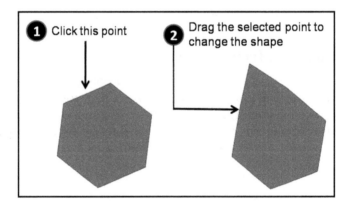

Similarly, we can delete a point by clicking on it and then hitting *Delete*. We can also delete multiple points by selecting multiple points by clicking and holding down the mouse button and then by hitting *Delete*.

Creating points from coordinate data

There is a lot of data with spatial components stored in spreadsheets and tables. One of the most common forms of tabular spatial data is *x* and *y* coordinates that are stored in a delimited text file. The data may have been collected with a GPS receiver, it may have been generated by a surveyor, or it may have been transcribed from topographic maps. Regardless, QGIS can map these coordinates as points by using the **Add Delimited Text Layer** tool . This tool can be found by navigating to **Layer | Add Layer | Add Delimited Text Layer**, or by using the **Manage Layers** toolbar.

Delimited text data is simply a table with column breaks that are identified by a specific character, such as a comma. With this tool, QGIS can accept either *x* and *y* coordinates or **well-known text (WKT)** representations of geometry. WKT can contain point, line, or polygon geometry. The following is some sample data, `cougar_sightings.csv`, viewed in a text editor. This is a comma-delimited file with *x* and *y* coordinate values:

```
"SAMPID,C,20","SEX,C,10","UTM_X,N,19,11","UTM_Y,N,19,11"
PA087,F,115556.044021,3486272.88304
PA097,F,116870.543644,3489102.55056
PA098,M,116148.894117,3483420.50411
PN001,M,482000.018751,3700998.34463
PN002,M,296192.720405,4053069.38808
PN003,M,347990.948523,3990302.26593
PN004,F,431049.74714,3998099.74491
PN005,F,498461.953615,4013066.46126
PN006,F,319083.556347,3988585.77826
```

In this example, the first row contains the column names and the definitions for the data type in each column. The column names and definitions are enclosed in quotes and are separated by commas. The first column reads `"SAMPID, C, 20"`. In this case, the field name is `SAMPID`. It is a text field signified by the letter `C`, which stands for *character*, with a width of `20` characters. The final two columns contain the coordinates. These are numeric fields signified by the `N` character. They have a precision of `19` and a scale of `11`.

QGIS has three requirements for the delimited text file to be mapped:

- The first row must be a delimited header row of field names
- The header row must contain field-type definitions
- If the geometry values are stored as x and y coordinate values, they must be stored as numeric fields

The **Create a Layer from a Delimited Text File** tool is simple but robust enough to handle many file-format contingencies. The following is the workflow for mapping data held in such a file:

1. Navigate to **Layer** | **Add Layer** | **Add Delimited Text Layer**.
2. Select the filename by clicking on **Browse...** and locate the delimited text file on your system. QGIS will attempt to parse the file with the most recently used delimiter.
3. Select **Layer name**. By default, this will be the prefix of the delimited text file.
4. Use the **File format** radio boxes to specify the format of the delimited text file. You will see how QGIS is parsing the file by the example at the bottom of the **Create a Layer from a Delimited Text File** window. The following are the options for **File format:**
    - Choose **CSV** if it is a standard comma-delimited file.
    - **Custom delimiters** can be checked to identify other delimiters used. The choices are **Comma, Tab, Space, Colon, Semicolon, or Other delimiters**.
    - Choose the **Regular expression delimiter** option if you wish to enter the regular expression for the delimiter. For example, \t is the regular expression for the tab character.
5. The **Record and Fields Options** section allows you to set different options for records and fields. Some of the options are described as follows:
    - We can set the Number of header lines to discard according to how many lines to disregard before beginning to read the data.
    - You can specify the number of header lines to discard. In most cases, this option will be set to **First record has field names**.
    - Check **Detect field types** to detect the type of data (this option is added in QGIS 3.6).

- Check **Trim fields** if you need to trim leading or trailing spaces from your data.
- Check **Discard empty fields** to prevent empty fields from being put into the output.
- If commas are also the separators for decimal-place values, check **Decimal separator is a comma**.

6. Once the file has been parsed, choose an appropriate value from the **Geometry definition** option:
     - If your file contains x and y coordinates, choose **Point coordinates** and identify the fields containing the x and y coordinates.
     - Choose **Well known text (WKT)**, if your file contains WKT geometries. For this option, you will also need to choose the field containing the WKT geometry definitions.
     - If the file does not contain any spatial information, choose **No geometry** and the table will be loaded simply as a table.
     - We can set the CRS here using **Geometry CRS**. In this case, select **EPSG:26913** by clicking on the far right button, and then by writing **EPSG:26913** and then selecting it and clicking **OK**.

7. Additionally, you can choose to enable the following options:
     - **Use spatial index**: Creates a spatial index
     - **Use subset index**: Creates a subset index
     - **Watch file**: This setting watches for changes to the file by other applications while QGIS is running

There is a setting that can affect the behavior of the **Coordinate Reference System Selector** for both new layers and layers that are loaded into QGIS without a defined CRS. By navigating to **Settings** | **Options** | **CRS**, you can choose how these situations are handled. The choices are **Prompt for CRS**, **Use Project CRS**, or **Use default CRS displayed below**. The default setting is **Prompt for CRS**. However, if you have this set to **Use project CRS** or **Use default CRS displayed below**, then you will not be prompted to define the CRS as described earlier.

The following screenshot shows an example of a completed **Create a Layer from a Delimited Text File** tool:

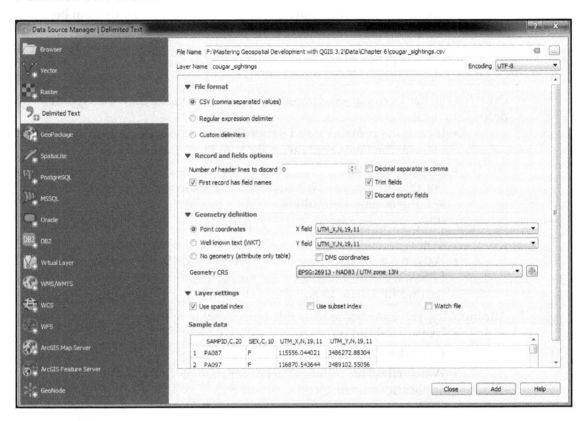

Once the tool has been run, a new point layer will be added to QGIS with all the attributes present in the original file (unless you chose to discard the empty fields). However, this is not a standalone GIS layer yet. It is simply a rendering of the tabular data within the QGIS project. As such, it will behave as any other layer. It can be used as an input for other tools, records can be selected, and it can be styled. However, it cannot be edited. To convert the layer to a standalone shapefile or another vector format, click on **Save as** under **Layer**, or right-click on the layer in the **Layers** panel and click on **Export | Save Features as**. Here, you can choose any OGR-supported file format, along with an output CRS of your choice. The cougar_sightings.csv sample data has coordinates in UTM zone 13 NAD83 or EPSG:26913.

The following screenshot shows the mapped data in the `cougar_sightings.csv` sample data:

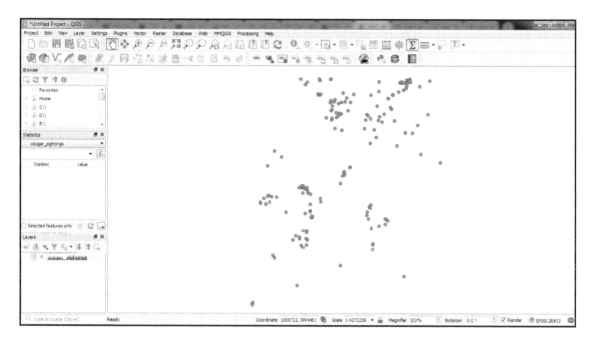

The sample data in `cougar_sightings.csv` is mapped by *x* and *y* coordinate values.

Mapping well-known text representations of geometry

As mentioned earlier, the **Add Delimited Text Layer** tool can also be used to map WKT representations of geometry. WKT can be used to represent simple geometries such as **Point**, **LineString**, and **Polygon**, along with **MultiPoint**, **MultiLineString**, and **MultiPolygon**. It can also represent more complex geometry types, such as geometry collections, 3D geometries, curves, triangular irregular networks, and polyhedral surfaces. WKT geometries use geometry primitives, such as **Point**, **LineString**, and **Polygon**, followed by the coordinates of vertices that are separated by commas.

For example, LINESTRING (30 10, 20 20, 40 30) would represent the line feature shown in the following diagram:

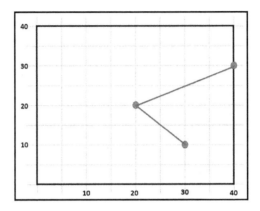

To demonstrate how WKT can be mapped via the **Add Delimited Text Layer** tool, we will map the Parcels_WKT.csv sample data file; this has WKT geometries for eight parcels (polygons):

1. Click on **Add Delimited Text Layer** by navigating to **Layer | Add Layer**.
2. Select the filename by clicking on **Browse...** and locate the delimited text file on your system. In this example, the Parcels_WKT.csv file is being used.
3. Choose an appropriate value for the **Layer name** field. By default, this will be the prefix of the delimited text file.
4. Use the **File format** radio buttons to specify the format of the delimited text file. This is a CSV file.
5. For **Record and fields options**, set the **Number of header lines to discard** option as **1**.
6. Set the **Geometry definition** option to **Well known text (WKT)**.
7. Click on ![icon] to the right of **Geometry CRS**. This will open a new dialog box, by means of which we can identify the coordinate reference system of the data. For this example, the data is in **EPSG:2903,** so we select **EPSG:2903** by typing it and selecting it and then clicking **OK**.

The following screenshot shows an example of a completed **Create a Layer from a Delimited Text File** tool set up to parse a WKT file:

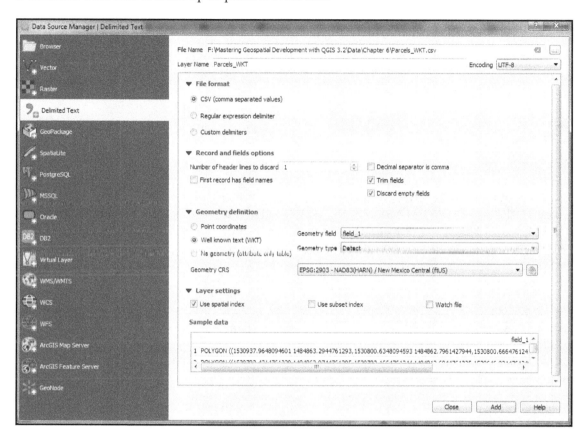

The data layer will be added to the **Layers** list and will behave as any other vector layer. The following diagram shows the resulting parcel boundaries:

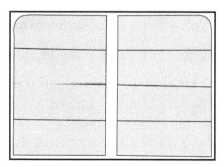

Well-known text representations of parcels are mapped via the **Add Delimited Text Layer** tool

> An easy way to explore WKT geometries is to use the get WKT plugin. This allows you to click on a selected feature (in the QGIS map canvas) and see the WKT for that feature. The WKT can be copied to the clipboard.

Geocoding address-based data

Another useful and common tabular spatial data source is a street address. There are many applications for geocoding addresses, such as mapping the customer base for a store, members of an organization, public health records, or incidences of crime. Once they are mapped, the points can be used in many ways to generate information. For example: they can be used as inputs to generate density surfaces, or they can be linked to parcels of land, and characterized by socio-economic data. They may also be an important component of a cadastral information system.

An address geocoding operation typically involves the tabular address data and a street network dataset. The street network needs to have attribute fields for address ranges on the left- and right-hand sides of each road segment. You can geocode within QGIS using a plugin named **MMQGIS** (`http://michaelminn.com/linux/mmqgis/`). **MMQGIS** is a collection of vector data processing tools developed by Michael Minn. This is preinstalled with QGIS 3.2.2, so you don't need to install it separately.

MMQGIS has many useful tools. For geocoding, you will use the tools found in **Geocode** under **MMQGIS**. There are three tools there: **Geocode CSV with Google/ OpenStreetMap**, **Geocode from Street Layer**, and **Street Address Join**. The first allows you to geocode a table of addresses using either the **Google Maps API** or the **OpenStreetMap/Nominatim** web service. This tool requires an internet connection but no local street network data. The web services provide the street network. The second tool uses a local street network dataset with address-range attributes to geocode the address data:

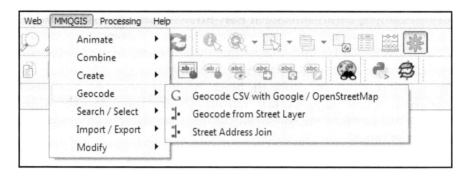

How address Geocoding works

The basic mechanics of address geocoding are straightforward. The street network GIS data layer has attribute columns containing address ranges on both the even and odd side of every street segment. In the following example, you can see a piece of the attribute table for the Streets.shp sample data. The columns **LEFTLOW**, **LEFTHIGH**, **RIGHTLOW**, and **RIGHTHIGH** contain the address ranges for each street segment:

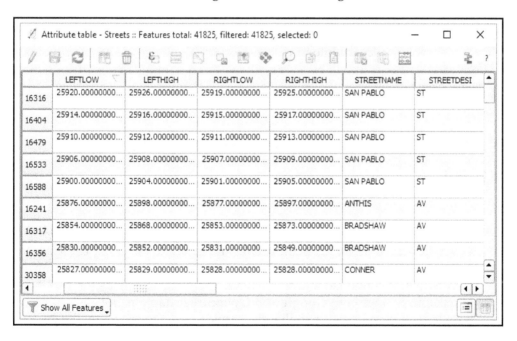

In the following example, we are looking at **Easy Street**. On the odd side of the street, the addresses range from **101** to **199**. On the even side, they range from **102** to **200**. If you want to map **150 Easy Street**, QGIS would assume that the address is located halfway down the even side of that block. Similarly, **175 Easy Street** would be on the odd side of the street, roughly three-quarters of the way down the block. Address geocoding assumes that the addresses are evenly spaced along a linear network. QGIS should place the address point very close to its actual position, but due to variability in lot sizes, not every address point will be perfectly positioned:

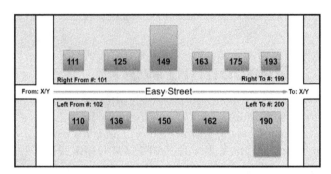

Now that you've learned the basics, you'll see how each MMQGIS geocoding tool works. Here, both tools will be demonstrated against the `Addresses.csv` sample data. The first example will use web services, while the second example will use the `Streets.shp` sample data. In both cases, the output will be a point shapefile containing all the attribute fields found in the source `Addresses.csv` files.

The first example – Geocoding using web services

Here are the steps for geocoding the `Addresses.csv` sample data using web services:

1. Load the `Addresses.csv` and `Streets.shp` sample data into QGIS Desktop.
2. Open the `Addresses.csv` sample data and examine the table. These are the addresses of municipal facilities. Notice that the street address (for example: **150 Easy Street**) is contained in a single field. There are also fields for the city, state, and country. Since both **Google** and **OpenStreetMap** are global services, it is wise to include such fields so that the services can narrow down the geography.
3. Navigate to **MMQGIS | Geocode | Geocode CSV with Google/OpenStreetMap**. The **Web Service Geocode** dialog window will open.

4. Select an appropriate value for the **Input CSV File (UTF-8)** field by clicking on **Browse...** and locating the delimited text file on your system.

5. Select the address fields by clicking on the drop-down menu and insert suitable values in the **Address Field**, **City Field**, **State Field**, and **Country Field** fields. MMQGIS may identify some or all of these fields by default if they are named with logical names such as `Address` or `State`.

6. Choose the web service, OpenStreet. If you want to use Google Map, then you have to get a Google API by observing the steps mentioned at `https://developers.google.com/places/web-service/get-api-key`.

7. Name the output shapefile by clicking on **Browse...**.

8. Select a value for the **Not Found Output List** field by clicking on **Browse...**. Any records that are not matched will be written to this file. This allows you to easily see and troubleshoot any unmapped records.

9. Click on **OK**. The status of the geocoding operation can be seen in the lower-left corner of QGIS. The word **Geocoding** will be displayed, followed by the number of records that have been processed.

10. The output will be a point shapefile and a CSV file listing the addresses that were not matched.

The following screenshot shows the completed **Web Service Geocode** tool:

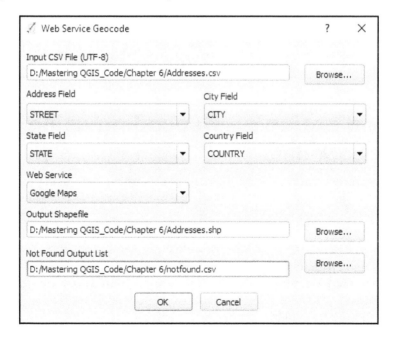

Two additional attribute columns will be added to the output address point shapefile: `addrtype` and `addrlocat`. These fields provide information regarding how the web geocoding service obtained the location. These may be useful for accuracy assessment.

`addrtype` is the `OpenStreetMap` class attribute or the `Google` `<type>` element. This will indicate what kind of address type was used by the web service (highway, locality, museum, neighborhood, park, place, premise, route, train station, university, and so on).

`addrlocat` is the `OpenStreetMap` type attribute, or the `Google` `<location_type>` element. This indicates the relationship of the coordinates to the feature addressed (approximate, geometric center, node, relation, rooftop, way interpolation, and so on).

If the web service returns more than one location for an address, the very first location that appears in the result will be used as the output feature.

To use this plugin, a strong internet connection is required. They place restrictions on the volume and number of addresses that act as a constraint on what addresses can be geocoded and for what time limit.

 Please visit the Google Geocoding API website, `https://developers.google.com/maps/documentation/geocoding/intro`, for more details.

Geocoding via these web services can be slow. If you don't get the desired results with one service, try the other.

The second example – Geocoding using local street network data

Here are the steps for geocoding the `Addresses.csv` sample data using local street network data:

1. Load the `Addresses.csv` and the `Streets.shp` sample data.
2. Open the `Addresses.csv` sample data and examine the table. This contains the addresses of municipal facilities. Notice that there is an address column (for example: **150 Easy Street**) along with separate columns for number (**150**) and street (**Easy**). This tool requires that the number and street address components be held in separate fields.
3. Install and enable the MMQGIS plugin.

4. Navigate to the **MMQGIS | Geocode | Geocode from Street Layer** menu and open the **Geocode from Street Layer** dialog window.

5. Select the `Addresses.csv` sample data as the **Input CSV File (UTF-8)** field by clicking on **Browse...** and locating the delimited text file on your system.

6. Select the street name field from the **Street Name Field** drop-down menu.

7. Select the number field from the **Number Field** drop-down menu.

8. Select the ZIP field from the **ZIP Field** drop-down menu.

9. Select the street GIS layer loaded in QGIS from the **Street Layer** drop-down menu.

10. Select the street name field of the street layer from the **Street Name Attribute** drop-down menu.

This tool allows geocoding from the street address ranges, or via **From X Attribute** and **To X Attribute** coordinates. The latter assumes that you have attribute columns with the To and From coordinates for each street segment. To geocode via the To and From coordinates, select the **From X Attribute**, **To X Attribute**, **From Y Attribute**, and **To Y Attribute** fields from the drop-down menu.

11. In this example, only address ranges will be used. Populate the **From X Attribute**, **To X Attribute**, **From Y Attribute**, and **To Y Attribute** drop-down menus with the **(street line start)**, **(street line end)**, **(street line start)**, and **(street line end)** options.

12. Since address ranges will be used for geocoding, select the **Left From Number**, **Left To Number**, **Right From Number**, and **Right To Number** attributes from the drop-down menu.

13. If the street data has left and right ZIP code attributes, select **Left Zip** and **Right Zip** from the drop-down menu. Since the `Streets.shp` sample data does not have ZIP code attributes, these options will be left blank **(none)**.

14. The **Bldg. Setback (map units)** option can be used to offset geocoded address points from the street centerline. This should represent how far buildings are from the middle of the street in map units. In this case, the map units are in feet. Enter a map unit value of **20**.

15. Name the output shapefile by clicking on the **Browse...** button.

Geocoding operations are rarely 100-percent successful. Street names in the street shapefile must match the street names in the CSV file exactly. Any discrepancy between the name of a street in the address table and the street attribute table will lower the geocoding success rate.

16. The tool will save a list of the unmatched records. Complete the **Not Found Output List** field by clicking on the **Browse...** button and name the comma delimited file. Any records that are not matched will be written to this file. This allows you to easily see and troubleshoot any unmapped records.

17. Click on **OK**.

18. The output will be a point shapefile and a CSV file listing the addresses that were not matched. In this example, the output shapefile will have 199 mapped address points. There will be four unmatched records described in the **Not Found CSV** list.

The following screenshot shows the completed **Geocode from Street Layer** tool:

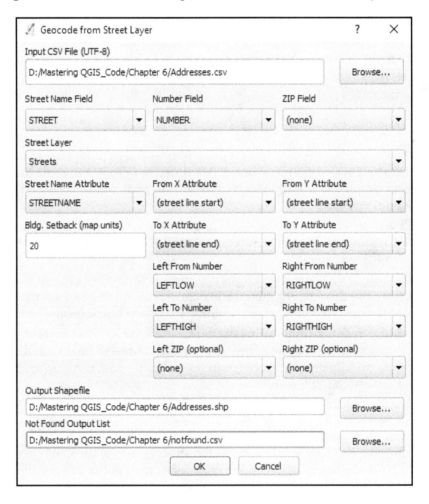

The following tip describes the typical workflow and issues often encountered in the geocoding process:

> Geocoding is often an iterative process. After the initial geocoding operation, review the **Not Found CSV** file. If it's empty, then all the records were matched. If it has records in it, compare them to the attributes of the streets layer. This will help you determine why the records were not mapped. It may be due to inconsistencies in the spellings of street names. It may also be due to a street centerline layer that is not as current as the addresses. Once the errors have been identified, they can be addressed by editing the data or by obtaining a different street centerline dataset. The geocoding operation can be rerun on the unmatched addresses. This process can be repeated until all the records are matched.
>
> Use the Identify tool ( ) to inspect the mapped points, and the roads, to ensure that the operation was successful. Never take a GIS operation for granted. Check your results with a critical eye.

The following screenshot shows the results of geocoding addresses via street address ranges. The addresses are shown with the street network used in the geocoding operation:

Georeferencing imagery

Maps and aerial photographs in hard copy have a lot of valuable data on them. When this data needs to be brought into a GIS, they are digitally scanned to produce raster imagery. The output of a digital scanner has a coordinate system, but it is a local coordinate system created by the scanning process. The scanned imagery needs to be georeferenced to a real-world coordinate system before it can be used in a GIS.

Georeferencing is the process of transforming the **coordinate reference system** (CRS) of a raster dataset into a new coordinate reference system. Often, the process transforms the CRS of a spatial dataset from a local coordinate system to a real-world coordinate system. Regardless of the coordinate systems involved, we'll call the coordinate system of the raster to be georeferenced **the source CRS,** and the coordinate system of the output **the destination CRS**. The transformation may involve shifting, rotating, skewing, or scaling the input raster from the source coordinates to the destination coordinates. Once a dataset has been georeferenced, it can be brought into a GIS and aligned with other layers.

Understanding ground control points

Georeferencing is done by identifying **ground control points** (GCP). These are locations on the input raster where the destination coordinate system is known. Ground control points can be identified in one of the following two ways:

- Using another dataset covering the same spatial extent that is in the destination coordinate system. This can be either a vector or a raster dataset. In this case, GCPs will be locations that can be identified on both datasets.
- Using datums or other locations with either printed coordinates or coordinates that can be looked up. In this case, the locations are identified and the target coordinates are entered.

Once a set of ground control points has been created, a transformation equation is developed and used to transform the raster from the source CRS to the destination CRS.

Ideally, GCPs are well distributed across the input raster. You should strive to create GCPs near the four corners of the image, plus several located in the middle of the image. This isn't always possible, but it will result in a better transformation.

Using the Georeferencer GDAL plugin

The **Georeferencer GDAL** plugin is a core QGIS plugin, meaning it will be installed by default. It is an implementation of the GDAL_Translate command-line utility. To enable it, navigate to **Plugins | Manage and Install Plugins,** and then click on the **Installed** tab and check the box to the left of the **Georeferencer GDAL** plugin:

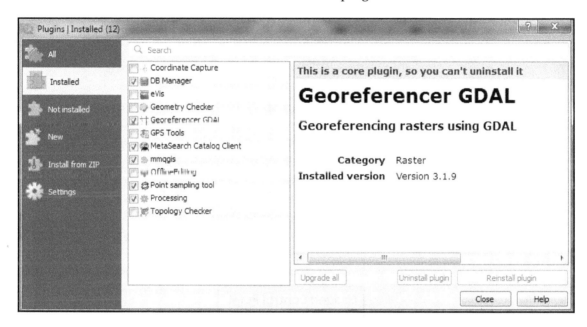

Once enabled, you can launch the plugin by clicking on **Georeferencer** under **Raster:**

The **Georeferencer** window has two main windows: the **Image Window** and the **Ground Control Point (GCP) Table** window. These windows are shown in the following screenshot:

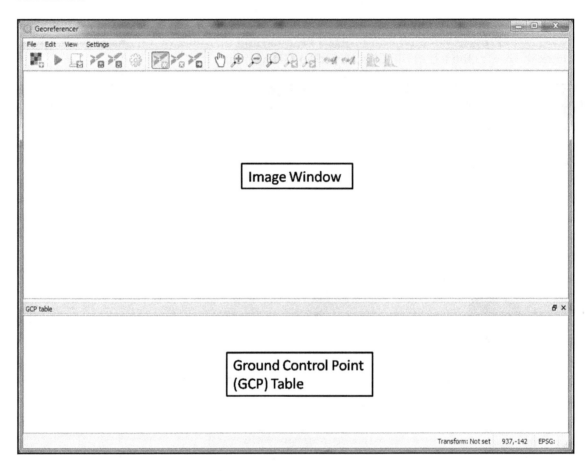

The general procedure for georeferencing an image is as follows:

1. Load the image to be georeferenced into the **Georeferencer** image window by clicking on **Open Raster** under **File**, or by using the Open raster button (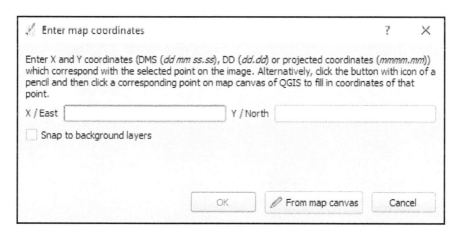).

2. If you are georeferencing the raster against another dataset, load the second dataset into the main QGIS Desktop map canvas.

   1. Begin to enter ground control points with the **Add point** tool (This tool, ，is also available via **Edit | Add point**). Regardless of which of the two ground control point scenarios you are working with, you need to click on the GCP point within the **Georeferencer** image window. Use the zoom and pan tools so that you can precisely click on the intended GCP location.

3. Once you click on the input raster with the **Add point** tool, the **Enter map coordinates** dialog box will open:

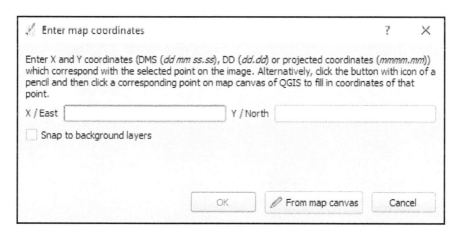

You are now halfway through entering the GCP. Follow the appropriate directions provided next. These will be different, depending on whether you are georeferencing against a second loaded dataset or against benchmarks, datums, or other printed coordinates on the input raster.

If you are georeferencing against a second loaded dataset, you need to follow these steps:

1. Click on the **From map canvas** button.
2. Locate the same GCP spatial location on the data loaded in the main QGIS Desktop map canvas. Click on that GCP location.
3. Use the zoom and pan tools so that you can precisely click on the intended GCP location.
4. If you need to zoom into the dataset within the QGIS Desktop map canvas, you will have to first zoom in and then click on the **From map canvas** button again to regain the **Add point** cursor and enter the GCP.
5. The **Enter map coordinates** dialog box will reappear with the **X / East** and **Y / North** coordinates entered for the point where you clicked on the QGIS Desktop map canvas:

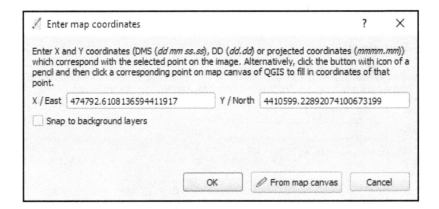

6. Click on **OK**.
7. As you enter GCP information for the source (**Source X/Source Y**) and destination (**Dst. X/Dst. Y**) coordinates, they will be displayed in the **GCP table** window in the **Georeferencer** window:

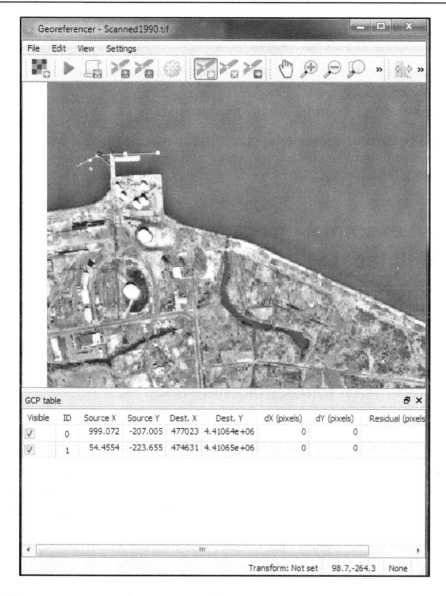

8. Repeat steps 1-4 and 1-7 to enter the remaining GCPs.

If you are georeferencing against benchmarks, datums, or other printed coordinates on the input raster, you need to perform the following steps:

1. Enter the **X / East** and **Y / North** coordinates in the appropriate boxes
2. Click on **OK**
3. Repeat steps 1-4 and 1-2 to enter the remaining GCPs

Now that you've learned the basic procedure for georeferencing in QGIS, we will go through two examples in greater detail. Here, you will learn all the options for using the **Georeferencer GDAL** plugin.

The first example – Georeferencing using a second dataset

In this example, the scanned1990.tif aerial photograph will be georeferenced by choosing ground control points from a more recent aerial photograph of the bridgeport_nj.sid area. The scanned1990.tif image is the result of scanning a hard copy of an aerial photograph. The bridgeport_nj.sid image file covers the same portion of the planet and is in the EPSG:26918 - NAD83 / UTM zone 18N CRS. Once the georeferencing operation is completed, a new copy of the scanned1990.tif image will be created in the EPSG:26918 - NAD83 / UTM zone 18N CRS.

Getting started

1. Launch QGIS Desktop and load the bridgeport_nj.sid file into the QGIS Desktop map canvas. (Note that you may need to navigate to **Properties |
Symbology** for this layer and set the **Min/max values settings** option to **Min / max** so that the image renders properly.)
2. Click on **Georeferencer** under **Raster** to launch the plugin. The **Georeferencer** window will open.

3. Load the `scanned1990.tif` aerial photograph into the **Georeferencer** image window by clicking on **Open Raster** under **File** or by using the open raster button,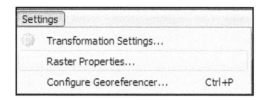

4. Click on **Settings** | **Raster Properties** to set CRS.

5. Now, click on for setting the CRS, shown as follows:

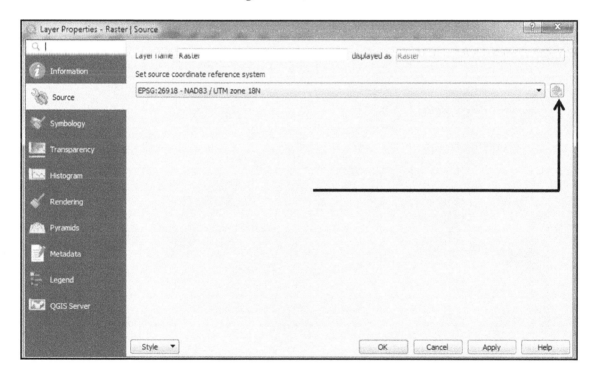

6. Now, type EPSG:26918 in **Filter,** select **EPSG:26918 - NAD83 /UTM zone 18N**, and click on **OK**. Then, click on **Apply** and **OK** in **Layer Properties**:

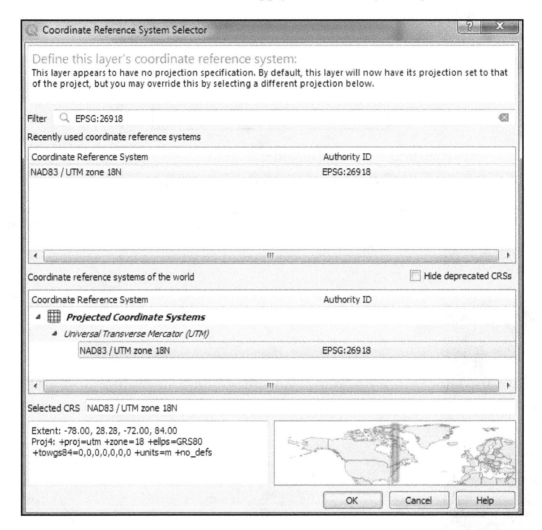

7. Arrange your desktop so that QGIS Desktop and the **Georeferencer** window are visible simultaneously.

8. Familiarize yourself with both datasets and look for potential GCPs. Look for precise locations such as piers, corners of roof tops, and street intersections.

Entering ground control points

Follow the steps mentioned here in order to enter ground control points.

1. Zoom into the area, using the zoom-in button ( ), where you will enter the first GCP in both the QGIS Desktop and the **Georeferencer** windows. Zooming in will allow you to be more precise.
2. Enter the first GCP into the **Georeferencer** image window using the **Add Point** tool ( ).
3. After clicking on the image in the **Georeferencer** image window, the **Enter map coordinates** window will open. Click on the **From map canvas** button. The entire **Georeferencer** window will momentarily disappear.
4. Click on the same location in QGIS Desktop.

If you have not first zoomed into the GCP area in QGIS Desktop, you can still do so. After zooming in, you will need to click on the **From map canvas** button, again or right-click to regain the **Add Point** cursor.

5. The **Enter map coordinates** window will reappear with the destination coordinates, populating the **X / East** and **Y / North** boxes. Click on **OK** to complete the **Ground Control Point**.
6. The **GCP table** in the **Georeferencer** window will now be populated with the source and destination coordinates for the first ground control point. The control point will also be indicated in both the **Georeferencer** image window and QGIS Desktop by a red dot.

If a GCP has not been precisely placed, the **Move GCP Point** tool ( ) can be used to adjust the position of the control point in either the **Georeferencer** image window or the QGIS Desktop map canvas.

GCPs can be deleted in two ways, as follows:

- Using the Delete point tool ( ) and clicking on the point in the **Georeferencer** image window
- By right-clicking on the point in the **GCP table** and choosing **Remove**

The following screenshot shows QGIS Desktop and the **Georeferencer** window with a single GCP entered:

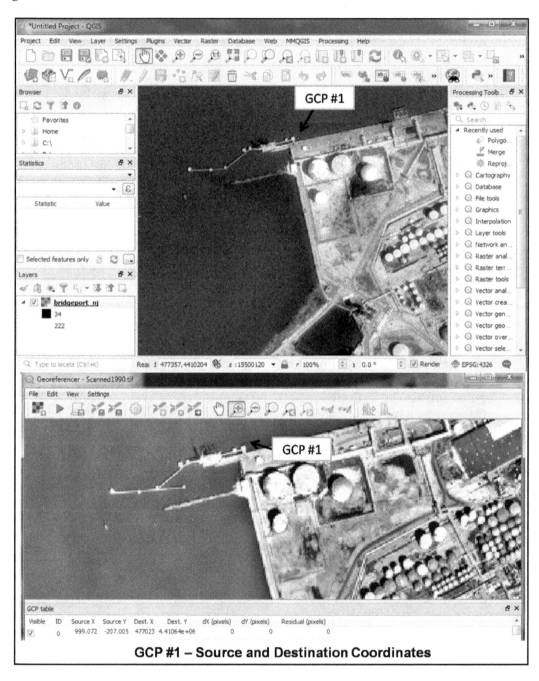

GCP #1 – Source and Destination Coordinates

Repeat steps 6 to 11 for the remaining ground control points until you have entered eight GCPs.

Use the pan and zoom controls to navigate around each image as required:

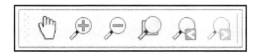

Once all the points have been entered, the **Save GCP Points As** button ( ) can be used to save the points to a text file with a `.points` extension. These can serve as part of the documentation of the georeferencing operation and can be reloaded with the **Load GCP Points** tool ( ) to redo the operation at a later date.

Transformation settings

The following steps describe how to set the transformation settings:

1. Once all the eight GCPs have been created, click on the transformation settings button ( ).
2. Here, you can choose appropriate values for the **Transformation type**, **Resampling method** fields, and other output settings. There are seven choices for **Transformation type**. This setting will determine how the ground control points are used to transform the image from the source to the destination coordinate space. Each will produce different results, and these are described as follows; for this example, choose **Polynomial 2**:

    *   **Linear:** This algorithm simply creates a world file for the raster and does not actually transform the raster. Therefore, this option is not sufficient for dealing with scanned images. It can be used on images that are already in a projected coordinate reference system, but that are lacking a world file. It requires a minimum of two GCPs.
    *   **Helmert:** This performs a simple scaling and rotational transformation. This option is only suitable if the transformation simply involves a change from one projected CRS to another. It requires a minimum of two GCPs.

- **Polynomial 1, Polynomial 2, Polynomial 3**: These are perhaps the most widely used transformation types. They are also commonly referred to as first- (affine), second-, and third-order transformations. The higher the transformation order, the more complex the distortion that can be corrected and the more computer power it requires. **Polynomial 1** requires a minimum of three GCPs. It is suitable for situations where the input raster needs to be stretched, scaled, and rotated. **Polynomial 2** or **Polynomial 3** should be used if the input raster needs to be bent or curved. **Polynomial 2** requires six GCPs, and **Polynomial 3** requires 10 GCPs.
- **Thin Plate Spline:** This transforms the raster in a way that allows for local deformations in the data. This may give similar results as a higher-order polynomial transformation, and is also suitable for scanned imagery. It requires only one GCP.
- **Projective**: This is useful for oblique imagery and some scanned maps. A minimum of four GCPs should be used. This is often a good choice when **Georeferencing** satellite imagery such as **Landsat** and **DigitalGlobe**.

There is no one best **Transformation type**. You may need to try several and then determine which generated the most accurate transformation for your particular dataset.

3. The following screenshot shows the setting of the **Transformation type** setting within the **Transformation settings** window:

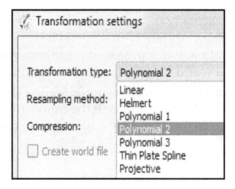

4. There are five choices for **Resampling method**. During the transformation, a new output raster will be generated. This setting will determine how the pixel values will be calculated in the output raster. Each is described here; for this example, choose **Linear**:

- **Nearest neighbour**: In this method, the value of an output pixel value will be determined by the value of the nearest cell in the input. This is the fastest method and it will not change pixel values during the transformation. It is recommended for categorical or integer data. If it is used with continuous data, it produces blocky output.

- **Linear:** This method uses the four nearest input cell values to determine the value of the output cell. The new cell value is a weighted average of the four input cell values. It produces smooth output because high and low input cell values may be eliminated in the output. It is recommended for continuous datasets. It should not be used on categorical data because the input categories may not be maintained in the output.

- **Cubic:** This is similar to **Linear**, but it uses the 16 nearest input cells to determine the output cell value. It is better at preserving edges, and the output is sharper than the linear resampling. It is often used with aerial photography or satellite imagery and is also recommended for continuous data. This should not be used for categorical data for the same reasons that were given for the linear resampling.

- **Cubic Spline:** This algorithm is based on a spline function and produces smooth output.

- **Lanczos**: This algorithm produces sharp output. It must be used with caution because it can result in output values that are both lower and higher than those in the input.

5. The following screenshot shows the **Resampling method** setting within the Transformation settings window:

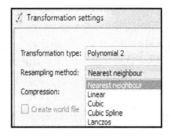

As is the case with **Transformation type**, there is no best **Resampling method**. Choosing the most appropriate algorithm depends on the nature of the data and how that data will be used after it has been georeferenced. **Nearest neighbour**, **Linear**, and **Cubic** are the most frequently used options.

6. Choose the target SRS by clicking on the **Browse** button. A window for choosing the target coordinate reference system will open. For this example, choose **EPSG: 26918**.

7. Name the output raster by clicking on the **Browse** button.

8. If an output map and output report are desired in PDF format, click on the browse button next to the **Generate pdf map** and **Generate pdf report** options and specify the output name for each. The report includes a summary of the transformation setting, GCPs used, and the root mean square error for the transformation.

9. Click on the **Set Target Resolution** box to activate the **Horizontal** and **Vertical** options for output pixel resolution in map units. For this example, leave this option unchecked.

10. Since raster data tends to be large, it may be desirable to choose a compression algorithm. There are four choices for **Compression**. Some choices offer better reductions in file size, while others offer better data access rates. For this example, use **None**. The choices are as follows:
    - **None:** This offers no compression.
    - **LZW:** This offers the best compromise between data access times and file size reduction.
    - **Packbits:** This offers the best data access times but the worst file size reduction.
    - **Deflate:** This offers the best file size reduction.

11. The **Use 0 for transparency when needed** option can be activated if pixels with a value of 0 should be transparent in the output. For this example, leave this option unchecked.

12. Click on **Load in QGIS** when this is done in order to have the output added to the QGIS Desktop map canvas.

> The Georeferencer tool can be configured by clicking on **Configure Georeferencer** under **Settings** in the **Georeferencing** window. Here, there are options for adding tips (labels) for the GCPS, choosing residual units, and specifying sheet sizes for the PDF report.

13. Click on **OK** to set the transformation settings:

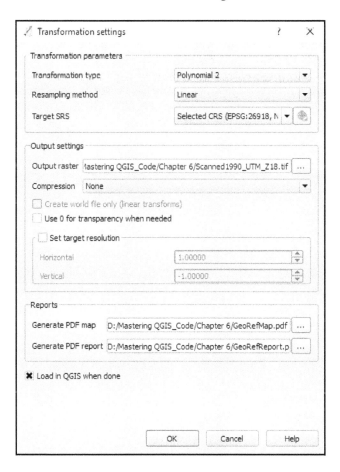

14. Once the **Transformation settings** values have been set, the `residual[pixels]` column in the **GCP table** will be populated. This column contains the **root mean square error (RMSE)** for each GCP. The mean RMSE for the transformation will be displayed at the bottom of the **Georeferencer** window.

The RMSE is a metric that indicates the quality of the transformation. It will change depending on the **Transformation type** value chosen. The general rule of thumb is that the RMSE should not be larger than half the pixel size of the raster in map units. However, it is only an indication. Another indication is how well the georeferenced imagery aligns with other datasets.

15. Additionally, the **Link Georeferencer to QGIS tool** ( ) and **Link QGIS to Georeferencer** tool ( ) will be activated. These will join the **Georeferencer** window to the QGIS Desktop map canvas, and they will be synched together when you use the pan and zoom tools.

The following screenshot shows the image in the **Georeferencer** window with eight GCPs entered. Their location is identified by numbered boxes within the image window. The **To** and **From** coordinates are displayed in the **GCP table** window, along with the RMSE values:

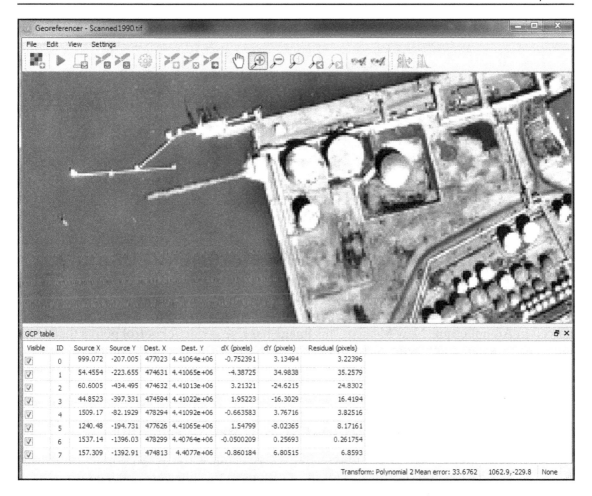

Visible	ID	Source X	Source Y	Dest. X	Dest. Y	dX (pixels)	dY (pixels)	Residual (pixels)
☑	0	999.072	-207.005	477023	4.41064e+06	-0.752391	3.13494	3.22396
☑	1	54.4554	-223.655	474631	4.41065e+06	-4.38725	34.9838	35.2579
☑	2	60.6005	-434.495	474632	4.41013e+06	3.21321	-24.6215	24.8302
☑	3	44.8523	-397.331	474594	4.41022e+06	1.95223	-16.3029	16.4194
☑	4	1509.17	-82.1929	478294	4.41092e+06	-0.663583	3.76716	3.82516
☑	5	1240.48	-194.731	477626	4.41065e+06	1.54799	-8.02365	8.17161
☑	6	1537.14	-1396.03	478299	4.40764e+06	-0.0500209	0.25693	0.261754
☑	7	157.309	-1392.91	474813	4.4077e+06	-0.860184	6.80515	6.8593

Transform: Polynomial 2 Mean error: 33.6762   1062.9,-229.8   None

Completing the operation

Click on the **Start Georeferencing** button (▶) button to complete the operation. The georeferenced raster will be written out and added to the QGIS Desktop map canvas if the option was checked in the **Transformation settings** window:

> The **Generate GDAL Script** tool (⬛) will write out the GDAL command-line syntax for the current georeferencing operation.

This code can be copied to the clipboard and used to write out to a file:

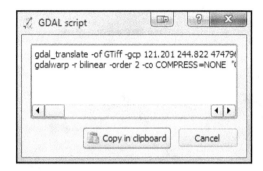

The second example – Georeferencing using a point file

This example will use the `zone_map.bmp` image. It displays zoning for a small group of parcels in Albuquerque, New Mexico. This image has five geodetic control points that are indicated by small points with labels (for example: I25 28). These control points are maintained by the **US National Geodetic Survey** (**NGS**). Here, you will learn how to load a precompiled set of ground control points to georeference the image:

1. Load the image into the **Georeferencer** window.
2. Choose **EPSG:2903 - NAD83(HARN) / New Mexico Central (ftUS)** for the CRS.
3. Click on the **Load GCP Points** tool ( ) and choose the `zone_map.points` file. The destination coordinates for the locations in this file were obtained from the NGS website (`http://www.ngs.noaa.gov/cgi-bin/datasheet.prl`). From the website, the **Station Name** link under **DATASHEETS** was used, and a station name search was conducted for each. The destination coordinates are in EPSG:2903 - NAD83(HARN) / New Mexico Central (ftUS).
4. The following diagram shows the image in the Georeferencer window with five GCPs entered based on datums whose destination coordinates were looked up online. Their location is identified by numbered boxes within the image window. The **To** and **From** coordinates are displayed in the **GCP table** window along with the RMSE values:

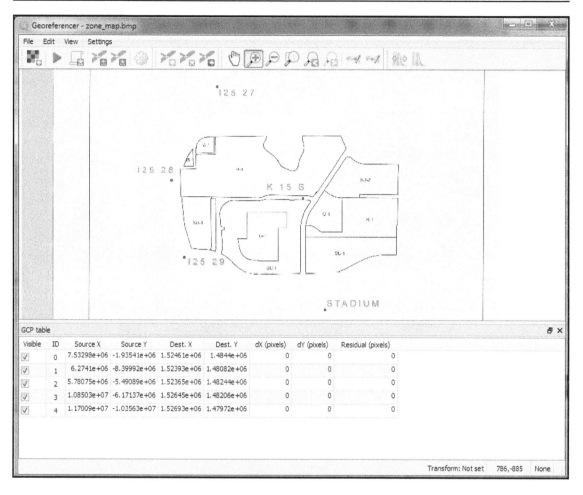

The `zone_map.points` file consists of five columns in a comma-delimited text file. The **MapX / MapY** columns are the destination coordinates. The **pixelX / pixelY** columns are the source coordinates, and the **enable** column has a value of 1 if the GCP is to be used in the transformation, and a value of 0 if it is not to be used. The following are the contents of this points file:

```
mapX,mapY,pixelX,pixelY,enable
1524608.32,1484404.47,7532975.55,-1935414.15,1
1523925.76,1480815.95,6274098.49,-8399918.00,1
1523645.13,1482436.21,5780754.77,-5490891.27,1
1526449.40,1482056.68,10850286.74,-6171365.35,1
1526925.10,1479718.37,11700879.35,-10356281.00,1
```

Click on **Transformation settings** under **Settings** and set the following parameters:

- Choose **Polynomial 1** as the **Transformation type**
- Choose **Nearest neighbour** as the **Resampling method**
- Choose **EPSG:2903** as the **Target SRS**
- Name the **Output raster**
- Choose a **Compression of None**
- Complete the **Generate PDF map** and **Generate PDF report** fields
- Check **Load in QGIS when done**
- Click on **OK**

The following screenshot shows the completed **Transformation settings**:

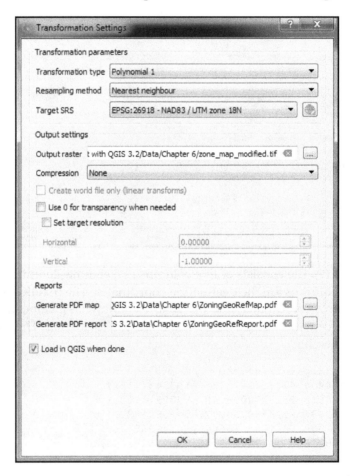

Click on the **Start Georeferencing** button (▶) to complete the operation. The georeferenced raster will be written out and added to the QGIS Desktop map canvas.

Checking the topology of vector data

In GIS, there are two main data models: **vector** and **raster**. They are called models because they are not real, but are representations of the real world. It is important that we ensure our data is modeling the world as accurately as possible. Vector datasets often have hundreds or thousands of features making it nearly impossible to verify each feature. However, using topology rules, we can let QGIS evaluate the geometry of our datasets and ensure that they are well constructed.

Topology is the relationship between contiguous or connected features in a GIS. Here, you will be introduced to the topology checker plugin. This plugin allows you to test topological relationships in your data and ensure that they are modeling the real world accurately. An example of a topological relationship rule is that polygons must not overlap. Imagine a country boundaries dataset. It is not possible for a point to be in two countries at once. Therefore, polygons in such a dataset should not overlap. The topology checker plugin can be used to test whether there are any overlapping polygons.

Installing the topology checker

Here are the steps for installing the topology checker plugin:

1. Navigate to **Plugins | Manage and Install Plugins** and click on the **All** tab.
2. In the search bar, type **topology**.
3. Select the **Topology Checker** plugin and click on **Close**.
4. Once enabled, the **Topology Checker** plugin can be found by navigating to **Vector | Topology Checker**.
5. When the **Topology Checker** window opens, it appears as a panel in QGIS Desktop.

Topological rules

Different sets of topological rules are available depending on the feature geometry: point, line, or polygon. Some rules test for relationships between features in a single layer, and some test the relationships between features of two separate layers. All participating layers need to be loaded into QGIS. The following topological rule tests are available.

Rules for point features

The rules for point features are as follows:

- **must be covered by**: This relationship test evaluates how a point layer interacts with a second vector layer. Points that do not intersect the second layer are flagged as errors.
- **must be covered by endpoints of**: This relationship test evaluates how a point layer interacts with a line layer. Points that do not intersect the endpoints of the second layer are flagged as errors.
- **must be inside**: This evaluates how a point layer interacts with a second polygon layer. Points not covered by the polygons are flagged as errors.
- **must not have duplicates**: This evaluates whether point features are stacked on top of one another. The additional points whose first point queried have the same x and y position (stacked) are flagged as errors.
- **must not have invalid geometries**: This checks whether the geometries are valid and if they are not, then it flags those features as errors.
- **must not have multi-part geometries**: This flags all multi-part points as errors.

Rules for line features

The rules for line features are as follows:

- **end points must be covered by**: This relationship test evaluates how a line layer interacts with a second point layer. The features that do not intersect the point layer are flagged as errors.
- **must not have dangles**: This test will flag features that are dangling arcs.
- **must not have duplicates:** This flags additional duplicate line segments (stacked) as errors.

- **must not have invalid geometries:** This checks whether the geometries are valid, and then it flags those features as errors.
- **must not have multi-part geometries:** This flags features that have a geometry type of multi-line as errors.
- **must not have pseudos:** This tests lines for the presence of pseudo nodes. This is when there is a pair of nodes where there should only be one. These can interfere with network analysis. The features with pseudo nodes will be flagged as errors.

Rules for polygon features

The rules for polygon features are as follows:

- **must contain:** This checks whether the target polygon layer contains at least one node or vertex from the second layer. If it doesn't, it is flagged as an error.
- **must not have duplicates:** This flags additional duplicated stacked polygons as errors.
- **must not have gaps:** This flags adjacent polygons with gaps as errors. Watersheds or parcel boundaries would be suitable for this test.
- **must not have invalid geometries:** This checks whether the geometries are valid. Some of the rules that define a valid geometry are as follows:
    - Polygon rings must close.
    - Rings that define holes should be inside rings that define exterior boundaries.
    - Rings should not self-intersect (they may neither touch nor cross one another).
    - Rings should not touch other rings, except at a point.
- **must not have multi-part geometries:** This flags all multi-part polygons as errors.
- **must not overlap:** This flags adjacent polygon features in the same layer that overlap one another as errors. Watersheds or parcel boundaries would be suitable for this test.
- **must not overlap with:** This relationship test evaluates how polygon features from the target layer interact with polygon features from a second polygon layer. Those that do will be flagged as errors.

Using the topology checker

The `parcels.shp` sample data will be used to demonstrate how to set up and test topological relationships. Here, the parcels polygon shapefile is loaded and the **Topology Checker** panel has been enabled by clicking on the **Topology Checker** button 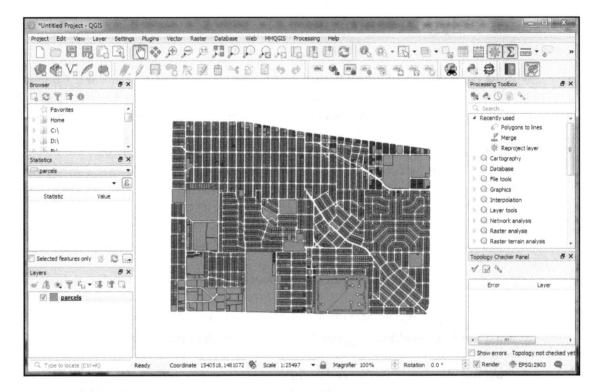 or **Topology Checker** under the **Vector** menu. The following screenshot shows the `parcels.shp` sample data loaded into QGIS desktop and the **Topology Checker** plugin enabled:

Here are the steps for configuring the **Topology Checker** plugin and evaluating the topology of the `parcels.shp` sample data:

1. Click on the **Configure** button in the **Topology Checker** panel to open the **Topology Rule Settings** dialog.
2. To set a rule, choose the target layer, **parcels**.

3. Next, choose the **must not have gaps** rule from the central drop-down menu:

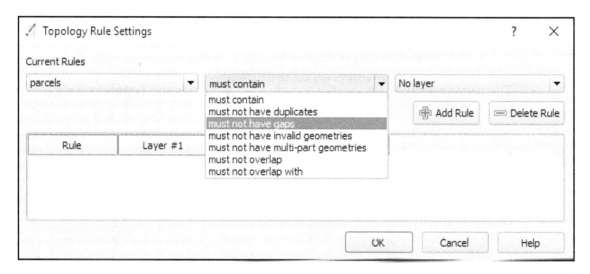

The list of available rules changes depending upon which target layer is chosen.

4. Since this rule involves only the target layer, the final drop for a second layer disappears.

5. Click on the **Add Rule** button (⊕ Add Rule).

6. Now, we will add a second rule. Set the target layer to **parcels** and the rule as **must not overlap**.

7. Click on the **Add Rule** button (⊕ Add Rule).

8. For the third rule, again set the target layer to **parcels** and set the rule as **must not have duplicates**.

9. Click on the **Add Rule** button.

10. For this example, we will test three rules against the **parcels** layer:

- **must not have gaps**
- **must not overlap**
- **must not have duplicates**

11. The rules that have been established are summarized in the **Topology Rule Settings** dialog box:

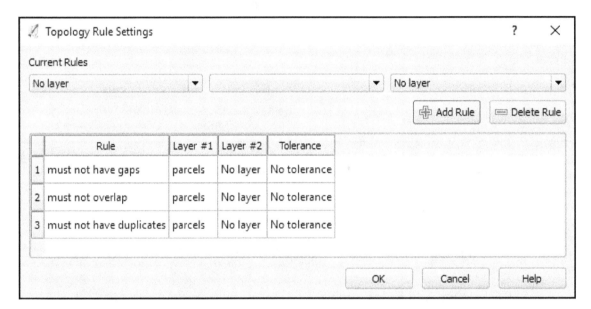

	Rule	Layer #1	Layer #2	Tolerance
1	must not have gaps	parcels	No layer	No tolerance
2	must not overlap	parcels	No layer	No tolerance
3	must not have duplicates	parcels	No layer	No tolerance

12. A rule can be deleted by selecting it and choosing **Delete Rule** (Delete Rule).
13. Click on **OK** when the topological rules have been defined.

As of version 0.1 of the **Topology Checker** plugin, the **Tolerance** setting is not operational. If it were, it would allow a tolerance to be set in map units. For example: you could test whether a bus stop's layer (point) is within 50 feet of the centerline of a road (line) by setting a tolerance of 50 feet and using the **must be covered by** rule.

14. Once the topology rules have been set, you can choose to validate the topology for the entire layer (**Validate All**) or just within the current map extent (**Validate Extent**).

15. For this example, choose **Validate All** ✓ .

16. The topology errors are displayed in the **Topology Checker Panel**. In this case, 17 errors are found, including six gaps, nine overlaps, and two duplicate geometries, as shown in the following screenshot:

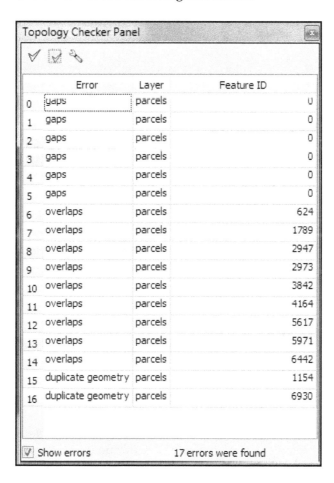

Now, we can see where the errors occur in the dataset. Errors will be highlighted in red, as shown in the following diagram:

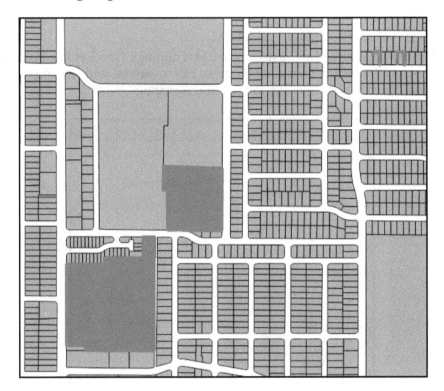

Repairing topological errors via topological editing

Once the geometry errors have been identified, the work of repairing the layer begins. In Chapter 1, *A Refreshing Look at QGIS*, we covered basic vector data editing that included layer-based snapping. In this final section, we will cover how to repair topological geometry errors via topological editing. We will continue to use the parcels.shp data as an example.

 Topological editing only works with polygon geometries

The editing approach taken depends on the topological error you are addressing. In the last section, three types of errors were found: gaps, overlaps, and duplicate geometries. These are three of the most common errors associated with polygon data, and we will look at how to repair these three types of errors.

Example 1 – Resolving duplicate geometries

Duplicate geometries are the most straightforward errors to address. Here are the step-by-step directions for resolving this type of topological conflict:

1. Toggle the editing option on for the **parcels** layer
2. Double-click on the first instance of a duplicate geometry in the **Topology Checker** error table to zoom to it
3. Use the **Select Features by Rectangle** tool ( ) to select the duplicate parcel
4. Open the **Attribute table**
5. Change the display filter in the lower-left corner to **Show Selected Features**:

6. Click on the **Delete selected features** ( )
7. Toggle off editing for this layer and save the changes

Example 2 – Repairing overlaps

To repair overlaps, there are some editing parameters with which you should familiarize yourself and that should be set.

Setting the editing parameters

Right-click on the toolbar and enable the snapping toolbar by clicking on it:

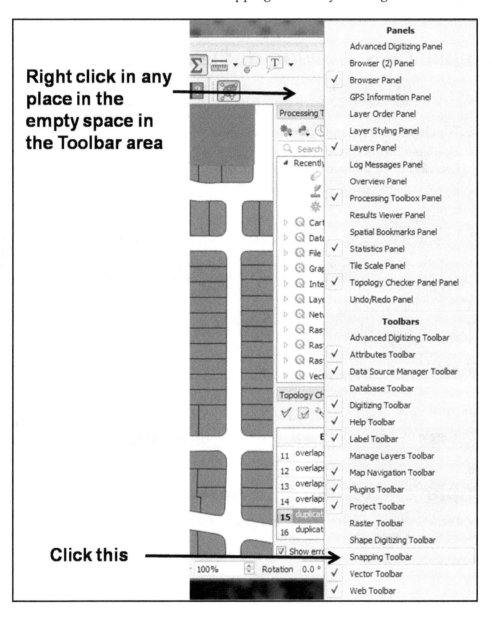

To activate the snapping toolbar, click on the magnet button as shown:

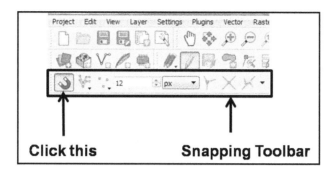

Click this   **Snapping Toolbar**

Click on **All Layers** ( ) and select **Active Layer** ( ). Choose **Vertex and segment** ( ) in the **Vertex** ( ) field and a value of **10 map units** in the **Snapping Tolerance in Defined Units** field.

Besides layer-based snapping options, you can also enable topological editing from the **Snapping Options** dialog box. Click on the **Enable topological editing** ( ) checkbox. Checking this option allows you to edit common boundaries in adjacent polygons. QGIS will detect shared polygon boundaries and vertices on these shared boundaries; they will only have to be moved once, and both polygons will be edited together.

There are two other editing options available here:

- **Enable snapping on intersection** ( ). This allows you to snap to an intersection of another background layer.

- **Enable Tracing (T)** ( ): using the offset tool on the snapping toolbar. You can trace around an object and offset your new feature from the traced feature by an amount specified by you.

For this example, leave these two options unchecked.

Click on **OK** to close the **Snapping options** window.

The **Snapping options** completed are displayed in the following screenshot:

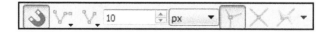

There are additional editing parameters that need to be set from **Options** under **Settings** on the **Digitizing** tab. Set the value for **Search radius for vertex edits** field to **10 pixels**, as shown in the following screenshot. Setting this value to something larger than zero helps to ensure that QGIS finds the correct vertex during an editing operation:

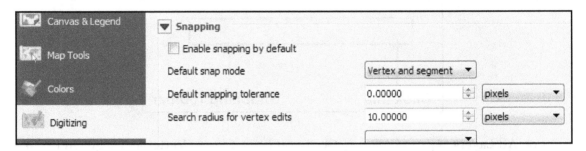

The following tip describes the use of transparency settings when editing overlapping polygons:

Adjusting the layer transparency can help when you work with overlaps. A 50-percent transparency settings will allow the overlaps to be visible.

Uncheck the **Show errors** option on the **Topology Checker** panel. This declutters the map canvas.

Here, we will work on the first overlap in the list that has a feature ID value of 624. To find this error, double-click on this record in the **Topology Checker** error table. QGIS will zoom into the location of the error shown in the following diagram. Here, we can see two parcels overlapping. The parcel on the right will be moved to the right to eliminate the overlap with the left-hand parcel:

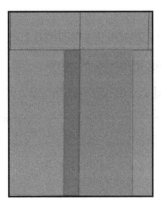

Repairing an overlap between polygons

Here are the step-by-step directions for repairing the overlaps found in the sample
`parcels.shp` data:

1. Toggle the editing option on for the parcels layer and select the polygon to edit
   with the **Select Feature by Rectangle** tool (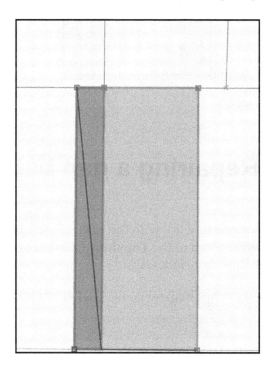). Each vertex will be displayed
   as a red graphic X.

2. The **Vertex tool** ( ) will be used to move the leftmost parcel and eliminate the
   overlap. It allows individual vertices to be moved.

3. Click on one of the parcel corners, and the vertices will appear as red boxes.

4. Click on the lower-left vertex of the right-hand side parcel to select it. It will turn
   blue.

5. Drag the selected vertex until it snaps to the boundary of the parcel it is
   overlapping. A blue line will appear showing the location of the virtual polygon
   boundary as you edit it, as shown in the following diagram:

This is the vertex of the right-hand overlapping parcel being moved.

6. Repeat the actions in the previous step for the vertex in the top-left corner.

7. Now, the vertices have been snapped to the boundary of the left-hand parcel and the overlap has been repaired. Click on **Validate Extent** 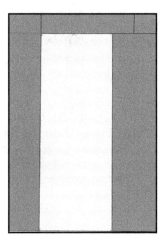 in the **Topology Checker** panel to ensure that the overlap has been solved. If so, no errors will be listed in the **Topology Checker** error table.

8. Any remaining overlaps can be fixed by repeating steps 1-4.

The repaired parcel is shown in the following diagram:

The overlap has now been repaired.

Example 3 – Repairing a gap between polygons

In this example, we will continue to work with the parcel.shp polygon layer. Here, we will focus on the first gap error listed in the **Topology Checker** error list that has a feature **ID** of 0. The steps to repair this are as follows:

1. Ensure that editing is still toggled on for the **parcels** layer.

2. Double-click on the error in the table so that QGIS will zoom to the area. You will see a small horizontal gap between two parcel polygons, as shown in the following diagram:

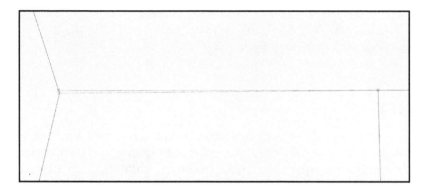

3. The same editing parameters that were set in Example 2 will be used here.
4. Zoom in a bit closer to the problem area.
5. Select the parcel to the north using the select feature by the rectangle tool.
6. Using the vertex tool, select the lower-left vertex and drag it to snap with the other two parcels and close the gap:

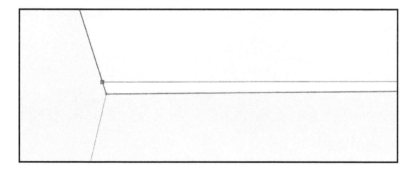

7. Verify that the issue has been resolved by clicking on **Validate Extent** in the **Topology Checker** panel.
8. The remaining gaps can be repaired using steps 1-7.

Toggle off editing for the parcels layer and save the edits.

 If you have too many topological errors to repair manually, you can import your data into a **GRASS** database. GRASS has a topological vector data model. The GRASS command, v.clean, will repair a lot of errors. The cleaned GRASS vector can then be exported into the file format of your choosing.

Summary

This chapter covered more advanced ways to create GIS data from different sources. We provided explanations and step-by-step examples of mapping raw coordinate data, geocoding address-based data, georeferencing imagery, validating vector data with topological rules, and topological editing. With the topics covered to this point, you will be able to work with a variety of vector, raster, and tabular input data.

In the next chapter, we will switch from preparing and editing data to visualizing spatial data. We will begin with live layer effects and use cases for inverted polygon shapeburst fills. We will then discuss how to render data with the **2.5D** renderer and how to create a true 3D scene with the **QGIS2ThreeJS** plugin. We will then cover a step-by-step description of creating a map series using the Atlas feature of the print composer, followed by a look at geometry generators and finally conclude by looking at **Data Plotly** plugin for drawing **D3** plots in **QGIS**.

Advanced Data Visualization

7

This chapter will provide readers with more advanced ways to visualize data. In QGIS 3.x, there are now new functionalities added for data visualization including 3D data visualization and animation (animation feature is added in QGIS 3.4). There are now many visualization tools that are unique to QGIS and allow you to extract information through beautiful data visualizations of both 2D and 3D. For example, digital elevation model can convey more information graphically if it is displayed in 3D view and summary data can be more meaningfully conveyed with box plot as we will cover in this chapter.

These are the topics that we will cover in this chapter:

- Using live layer effects
- Creating beautiful effects with inverted polygon shapeburst fills
- Using the 2.5D renderer
- Creating 3D views
- Creating an Atlas
- The power of geometry generators
- Working with the Data Plotly plugin

Using live layer effects

You can make stunning visuals using live layer effects. From the **Style** tab of **Layer Properties**, you can enable this feature by clicking **Draw effects** in the **Layer Rendering** section, as seen here:

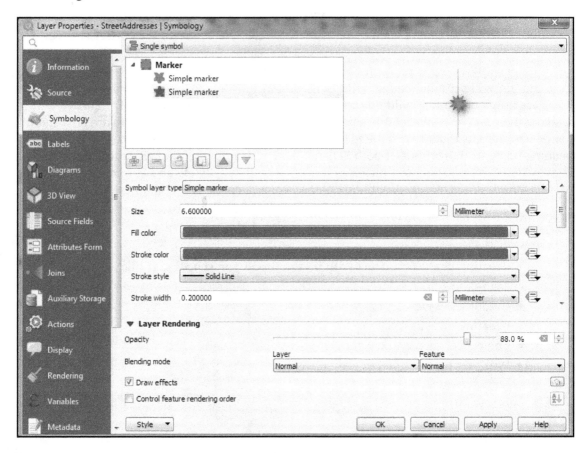

Once the **Draw effects** feature has been enabled, the **Customize effects** button is also enabled.

Click the **Customize effects yellow star** button to open the **Effect Properties** window shown here:

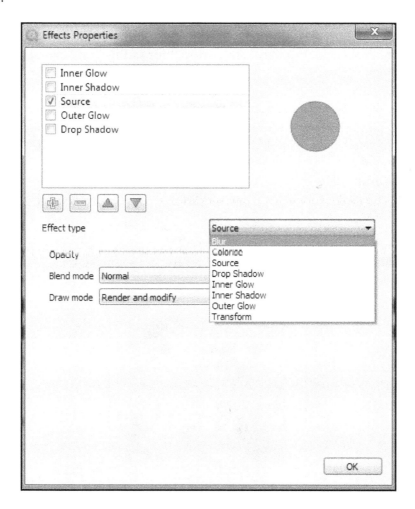

The **Effect type** dropdown gives you access to eight different effects. For most of these, transparency, blending modes, and draw modes are customizable. The available effects are as follows:

- **Blur**: Creates a blurred version of the feature.
- **Colorise**: Allows you to change the color of the feature using brightness, contrast, and saturation sliders along with transparency, blending modes, and draw modes.

- **Source**: The original feature style.
- **Drop Shadow**: Creates a drop shadow around a feature. The sun angle (offset), shadow length, blur radius, transparency, and color are adjustable.
- **Inner Glow**: Creates a glow effect from the features edge inward along a customizable color gradient. The spread, blur radius, transparency, and color are adjustable.
- **Inner Shadow**: Creates a shadow effect from the features edge inward. The sun angle (offset), shadow length, blur radius, transparency, and color are adjustable.
- **Outer Glow**: Creates a glow effect from the features edge outward along a customizable color gradient. The spread, blur radius, transparency, and color are adjustable.
- **Transform**: Allows you to transform the feature symbol in a variety of ways: shear, scale, rotation, and *XY* translation.

The following examples will illustrate several of these effects using the `LiveLayerEffects.qgs` map file. The effects will be applied to the `Food Dessert` polygons and will show one particular dessert for illustrative purposes. Here is the original data without any live layer effects applied:

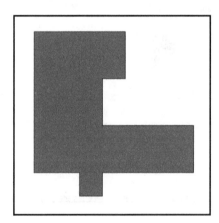

This is the drop shadow effect, which gives more visual weight to a feature and lifts it off the map a bit:

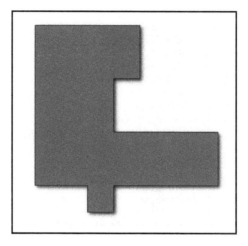

This is the **Blur** effect, which gives the polygon a fuzzy boundary:

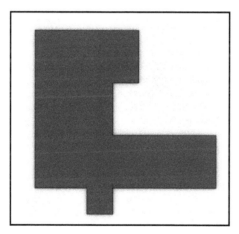

This is the inner shadow effect. With the default effect configuration, it creates a shadow on the interior of the upper-left sides:

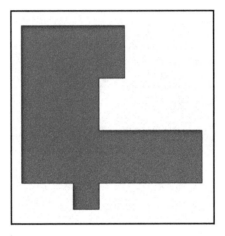

This is the **Outer Glow** effect. The map background was changed so the effect would be visible against a darker background:

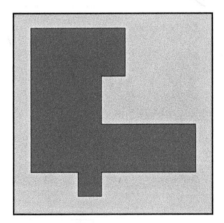

These effects can also be used in combination by enabling multiple effects in the **Effects Properties** window. These effects, along with blending modes, allow you to create cartographic effects available in no other GIS software! They are extremely powerful cartographic tools, and we encourage you to experiment with them.

Creating beautiful effects with inverted polygon shapeburst fills

The inverted polygon renderer can be used in combination with a shapeburst fill to create some beautiful effects. Next, you will see two examples of this in a single map covering Baltimore City, Maryland. You will begin by opening the `Baltimore.qgs` project. The data focuses on diabetes in Baltimore City neighborhoods:

Creating coastal vignettes

It's already a fairly nice map, but there is no data representing the harbor and Chesapeake Bay. It is simply a white hole in the data layers. To make it blue, you could click on **Project | Project Properties** and change the background color of the map to blue. However, let's learn a more powerful way to represent this feature:

1. Right-click on the MD_County_boundaries layer and choose **Duplicate**. A second copy of the layer will appear in **Layers Panel**.
2. Drag the new copy so that it is between the Neighborhood_diabetes and Baltimore City Boundary layers in the **Layers Panel**, and turn the layer on.

   We will now learn about inverted polygon styling, which is a way to style data outside the boundaries of a layer. It can be used to create a mask around a feature. Here, we will use it to represent Chesapeake Bay.

3. Open the **Layer Properties | Style** tab for the copied layer.
4. From the renderer dropdown, choose **Inverted Polygons**.
5. Select **Simple fill**.
6. Select a **Symbol layer type** of **Shapeburst fill**.
7. Under **Gradient colors**, you will be using the **Two color** method.
8. Color number one will have an RGB value of 225/255/255, which is a light blue.
9. Color number two will have an RGB value of 166/206/227, a darker blue.
10. Click **Apply** but do not close the **Layer Properties** window.

    The area beyond the extent of the county boundaries, which is the equivalent of Chesapeake Bay, is now styled with a light blue to darker blue colored ramp. It also uses the **Shapeburst fill** style, which fills the area with a gradient based on the distance from the edge.

11. Now, under **Shading style**, select **Shade to a set distance** and set the distance to 12.
12. Again click **Apply,** but do not close the **Layer Properties** window. This gives a little more definition to the center of the channel.

13. Finally, you will create a shoreline. Click the add symbol layer button ⊕.
14. Change from a **Symbol layer type** of **Simple fill** to **Outline: Simple line**. Give the line an RGB color of 31/120/180, which is a dark blue. You can create great effects by using composite renderers such as this.
15. Click **OK**.

The following is the **Layer Properties** window that you might see now:

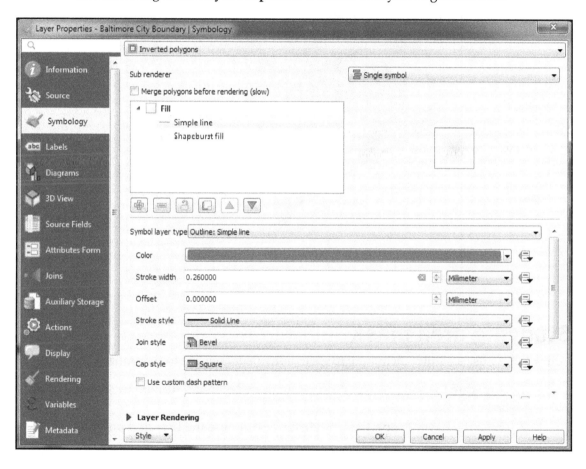

16. You have symbolized Chesapeake Bay, without having a layer representing water, using an inverted polygon shapeburst fill. Look at the following screenshot:

Studying area mask

You will now use a similar technique to give more visual weight to Baltimore City:

1. Duplicate the `Baltimore_City_Boundary` layer and turn on the newly copied layer.
2. Open the **Layer Properties | Style** tab and choose the **Inverted polygon** renderer.
3. Select **Simple fill** and choose the **Symbol layer type** of **Shapeburst fill**.

4. Under **Gradient colors**, you will again be using the **Two color** method.

5. Color number one will have an RGB value of 135/135/135, which is gray. Color number two will be white.

6. Under **Shading style**, select **Shade to a set distance** and set the distance to 4.

7. Under **Layer rendering**, slide the transparency slider so that **Layer transparency** is **45**.

8. Click **OK**.

You can see these settings here:

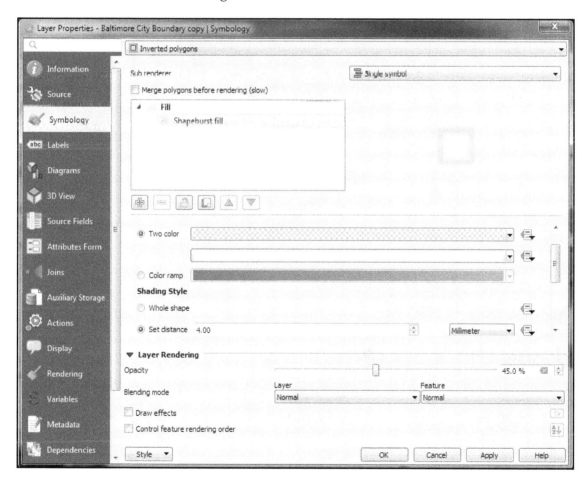

This styled the area beyond the city limits. You used a gray-to-white color ramp with a distance of 4, which creates a mask around the city boundary. The area beyond Baltimore City is underneath the white mask created by this styling so that visually, it falls into the background. The layer is also semi-transparent, so you can still see the underlying data. It creates an effect where the area of interest pops off the map and receives more visual weight:

Using the 2.5D renderer

QGIS now has a 2.5D renderer, so named because it doesn't actually allow you to generate a three-dimensional scene, but it does allow data to be rendered in such a way that height and some 3D perspective is shown. The most obvious use case for this renderer is showing building footprint polygons extruded upward, based on the building height. For this example, we are using the `lower_manhatten_buildings.shp` shapefile. This represents building footprints below 14th Street.

Importantly, it includes a NUM_FLOORS attribute, which will be used to extrude the buildings:

1. Open the **Layer Properties** | **Style** tab for this layer.
2. Choose the **2.5D** renderer.
3. For **Height,** you can use any column representing heights. Here, the NUM_FLOORS column is being used in an expression: "NUM_FLOORS" * 4. Using this multiplier gives extra extrusion. If no such attribute exists, you can simply use a numeric height for all features. The default is 10.
4. Keep the default **Roof** and **Wall** colors.
5. The **2.5D** expression is seen in the **Control feature rendering order** section.
6. Click **OK**:

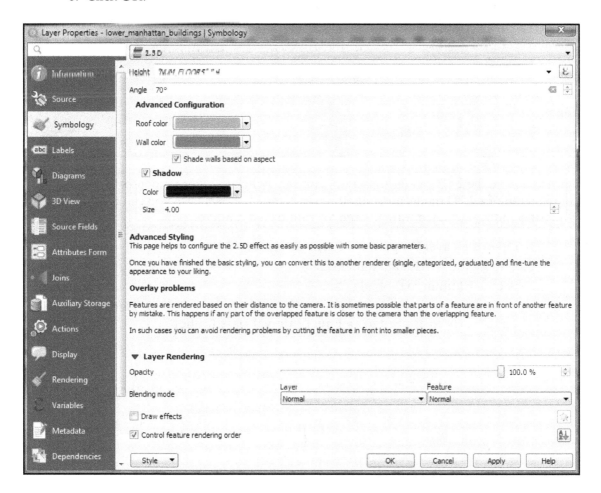

The buildings are now rendered with 2.5D representations of their height. In the following example, the `QuickMapServices` plugin is also used to add an OSM Standard basemap:

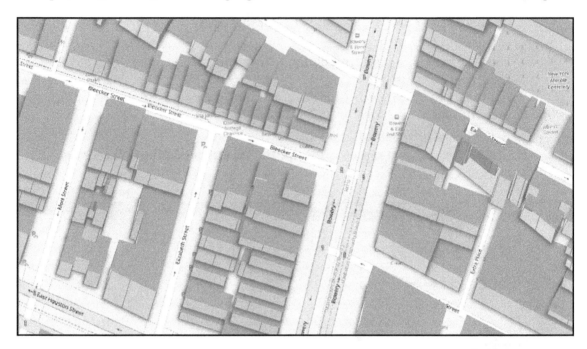

Once you have the 2.5D rendering set up, you can switch to other renderers to create a combination of a 2.5D rendering and other renderers. For example, you can combine a 2.5D rendering and a **graduated**, **categorized**, or **rule-based** renderer:

1. Open the **Layer Properties** | **Style** tab again.
2. Choose the **Graduated** renderer.
3. Set the **Column** to NUM_FLOORS.
4. Choose a **Color ramp** and classification scheme of your liking. Here, a blue color ramp has been applied to an **Equal Interval** classification.

5. You will notice that the **Control feature rendering order** section stays, as you set it up with the 2.5D renderer. This feature can be used without the 2.5D renderer to specify the rendering order of features. However, it is used by the 2.5D renderer to specify the distance from the camera.

6. Click **OK**.

Combining rederers will be as follows:

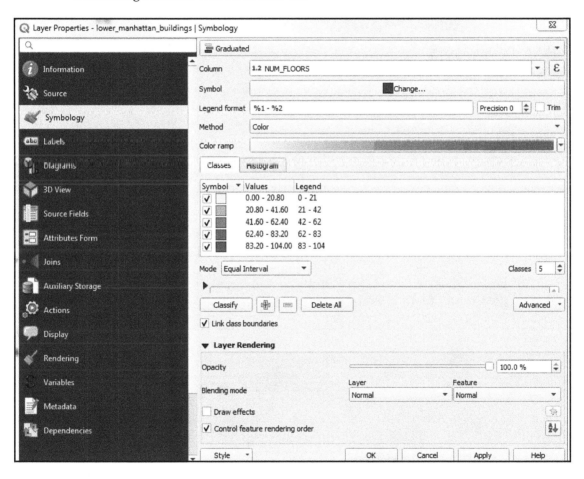

7. Now the buildings are extruded into 2.5D space and are styled according to their height:

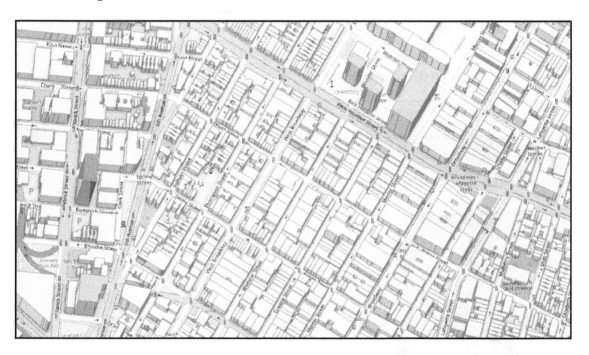

8. From the **Layer Properties | Style** tab, the ellipsis button to the right of the **Control feature rendering order** section can be used to edit the expression:

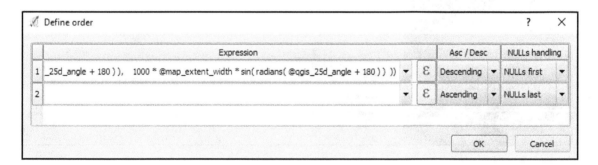

Creating 3D views

Starting from QGIS 3, QGIS now has its own 3D views capable and built-in instead of resorting to the assistance of any plugins. You can create 3D views by clicking on **View** | **New 3D Map View**.

Here, we will use the QGIS `terrain.qgs` project to generate a view of the Sandia Mountains outside Albuquerque, NM. The project has two layers: a styled **Digital Elevation Model** (**DEM**), and a hillshade layer. The hillshade layer has been set up with a multiply blending mode to create a color hillshade effect.

To generate the 3D view, follow these steps:

1. From the menu bar, select **View** | **New 3D Map Layer**.
2. Now, we will see that a 3D map of this is generated inside a new window as follows:

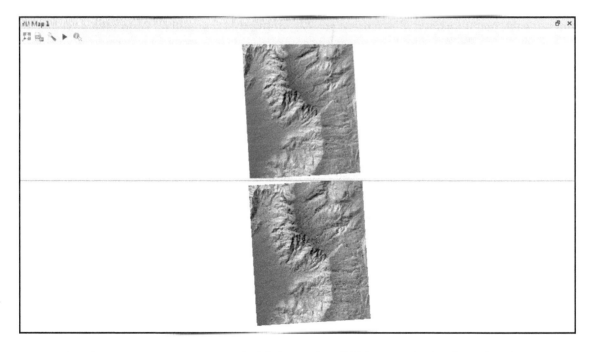

3. Now, to see it in 3D view, hold down the *Shift* key on the keyboard and move your mouse left or right to rotate this. Similarly, by holding down the *Shift* key and moving the mouse upward or downward, you can tilt it upward or downward.

Creating an Atlas

Often, you will need to create a series of maps, all with the same data; for example, street atlases for urban areas, or maps of watersheds in a national forest. The Atlas function in the Print Composer allows us to set up and configure a series of maps. Once configured, QGIS will autogenerate each map image. This can be a tremendous time saver and can also reduce human error.

Typically, map elements, such as the title or legend, will need to change with each atlas image. These map elements can be configured with Atlas to allow for map-by-map customizations. The key element in an Atlas is the **coverage layer**, whose features define the extent of each map and the number of maps.

Basic Atlas configuration

For this example, you will use the `BaltimoreAtlas.qgs` project. Our goal is to generate a map of each neighborhood. Since there are 271, doing this manually would take days. However, using the Atlas feature, we can accomplish this in a much shorter period of time.

To begin, open the project file. This contains the same data used earlier in this chapter, but the base layer has been replaced by a QuickMapServices Google Road basemap. The basic Atlas setup is done using the following steps:

1. Click **Project** | **New Print Layout...**.
2. Give the print layout the title **Baltimore**.
3. From the **Atlas** menu, choose **Atlas Settings**.
4. A new **Atlas generation** tab opens. Click **Generate an atlas**.
5. Set **Coverage layer** to **Baltimore Neighborhood** and set **Page name** to **Page**. If your coverage layer is one that you do not want to appear on the maps, click the **Hidden coverage layer** option. You can use expressions in the **Filter with** option to refine your coverage layer. You can also adjust the image generation order via the **Sort by** option:

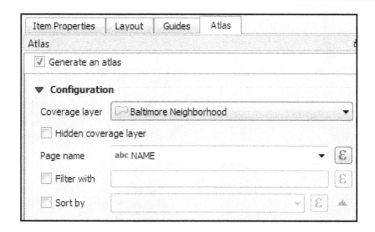

6. Click on  and then drag and create a rectangle in QGIS composer to get the map inside.

7. Now, switch to the **Item Properties** tab.

8. Click the **Controlled by Atlas** box. This control has three options:

   - **Margin around feature**: This zooms to every **Coverage** layer feature with a 10% padding to the spatial extent.
   - **Predefined scale**: This chooses the most appropriate scale from those set in **Settings | Options**.
   - **Fixed scale**: All maps will be rendered using the specified scale.

9. For this example, we will use the **Margin around feature** option.

10. Switch back to the **Atlas** generation tab.

11. Now, you can configure the outputs. You can set an expression to define the naming of each image. The output can be a single image per **Coverage** layer, or you can produce a multi-page PDF by checking the **Single file export when possible** option. This option is only utilized if you choose PDF as your output file format.

12. Once the basic setup is complete, you can use the **Preview Atlas** button on the Atlas toolbar to preview individual map images:

13. Use the **Previous** and **Next feature** buttons, along with the page number selector, to preview specific map pages. Uncheck the **Preview Atlas** button when done:

Dynamic titles

When creating a map series, you really want each map to have a custom title. Fortunately, it is possible to use variables to generate dynamic titles for each page. To do this, create a title text label with the **Add new label** tool. Scroll to the bottom of the **Item properties** tab to the **Variables** section.

There are a series of **Atlas variables** that can be leveraged to insert dynamic text:

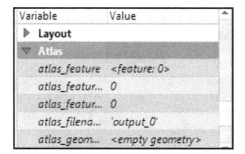

Now, click on **Add Item**, and then on **Add Legend**. Drag and create a rectangle that will hold the title for different maps. Insert an expression, expand **Variable**, double-click on the atlas_pagename variable to insert that variable into the expression, and then click **OK**. The variable is inserted as the title text.

Now, click on **Add Item** and then on **Add Legend**. Drag and create a rectangle that will hold the title for different maps. Insert an expression, expand **Variable**, and double-click on the atlas_pagename variable to insert that variable into the expression, then and click **OK**. The variable is inserted as the title text.

This is yet another great application of QGIS variables! Change the font size and font color as appropriate:

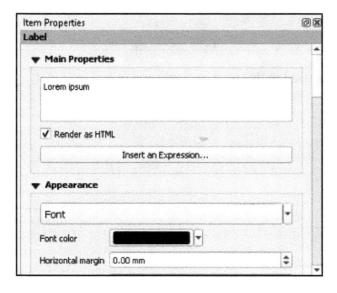

Using the **Preview Atlas** button, you can now see the title variable in action. The title will update to the coverage layer neighborhood for each page:

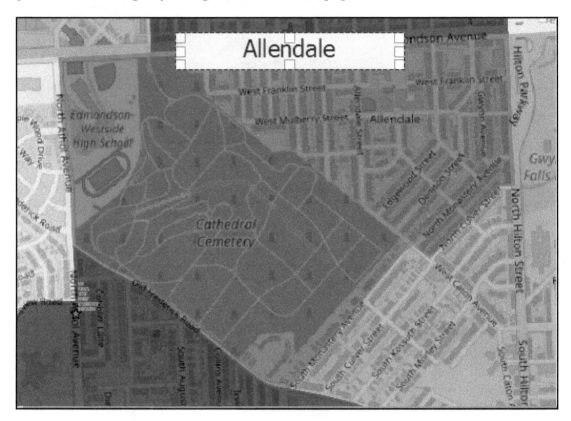

Atlas variables can be used in this way to create any other desired dynamic text on the composition.

Dynamic legends

Legend content can also be dynamic. After adding a legend, select **Item properties**. In the **Legend items** section, select the **Only show items inside current atlas feature** option. This will filter the legend based on the features visible on a given page:

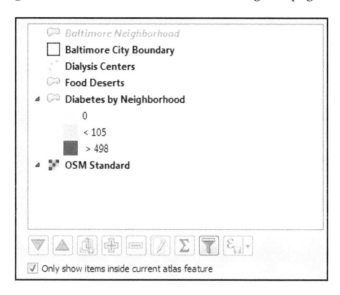

Highlighting the coverage feature

It is also possible to filter the contents of a layer to equal the coverage feature. In this example, the **Diabetes by Neighborhood** layer has been duplicated and given a single symbol of hollow yellow polygons. The renderer is then switched to **Rule-based**. The rule is set to `@atlas_featureid = $id`.

This expression filters the layer so that it is equal to just the feature whose ID equals that of the current Atlas coverage feature. In this example, the polygon is also given a drop shadow via live layer effects:

- Select the map object and switch to the **Item properties** tab.
- Click the **Controlled by Atlas** box. This control has three options:
  - **Margin around feature**: This zooms to every **Coverage** layer feature with a 10% padding around the spatial extent.
  - **Predefined scale**: This chooses the most appropriate scale from those set in **Settings | Options.**
  - **Fixed scale**: All maps will be rendered using the specified scale.
- For this example, we will use the **Margin around feature** option.
- Switch back to the **Atlas generation** tab.

Now, you can configure the outputs. You can set an expression to define the naming of each image. The output can be a single image per **Coverage layer,** or you can produce a multi-page PDF by checking the **Single file export when possible** option. This option is only utilized if you choose PDF as your output file format.

Once the basic setup is complete, you can use the **Preview Atlas** button on the **Atlas** toolbar to preview individual map images:

Use the **Previous** and **Next feature** buttons along with the page number selector to preview specific map pages. Uncheck the **Preview Atlas** button when finished:

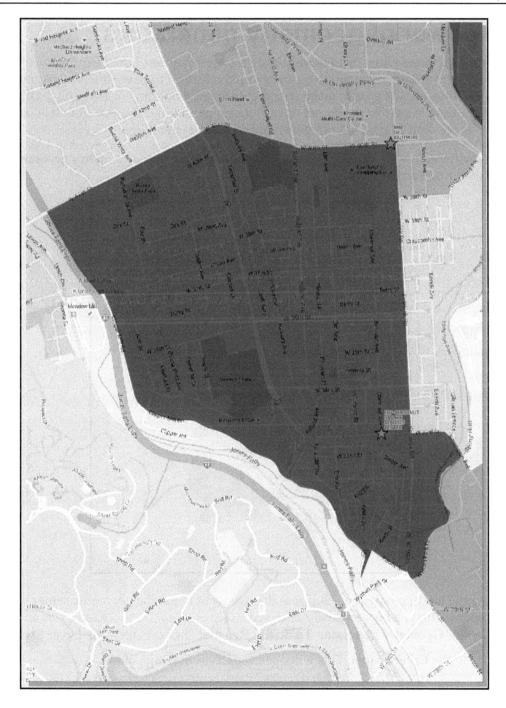

The power of geometry generators

Geometry generators allow us to modify and create geometries. Using expression engines, they allow us to modify or generate geometry on the fly and to have different geometries stacked together, allowing us to make creating complex rendering an easy and feasible task. They can be used to perform different operations on point, line, and polygon data, and using the expression engine, we can generate advanced symbology.

Now, we will use the `dhaka_gazipur.shp` file to see some of the geometry generators you can use with polygon data:

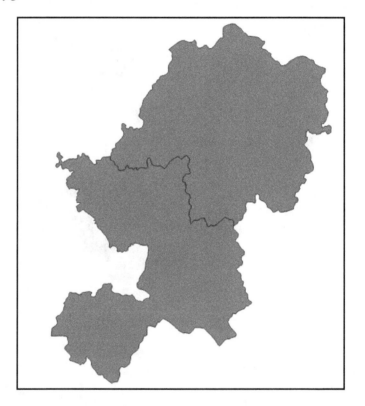

This shapefile has two polygons. Suppose that we want to get the geometric center of these two polygons. This can be very easily achieved using `centroid($geometry)` as an expression for **Geometry generator**. To do this, we first click on **Open the Layer Styling panel** in the **Layers** panel. This will open the **Layer Styling** panel.

Now, we will select **Simple fill** and then select **Geometry generator** under **Symbol layer type**:

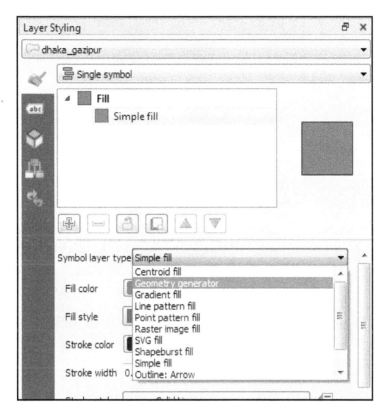

Now, select **Point/Multipoint** as the Geometry type and write `centroid($geometry)`:

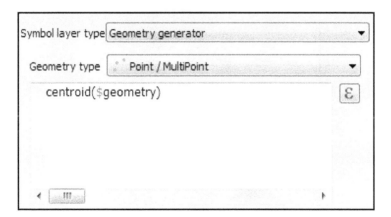

This gives us the centroid of these two polygons as shown:

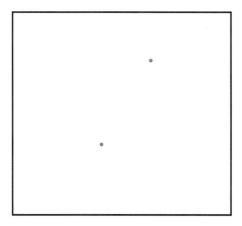

You can easily create a buffer around the shape file using **buffer()**. Suppose you want a 600-meter buffer around this polygon; write buffer($geometry, 600) as follows:

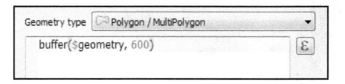

Now, you will get a buffer around it:

There are many other geometry generators available to use for conducting more complicated operations. You can access these by clicking on , and the **Expression** dialog will open.

Now you can see different functions under different categories. If you want to have an idea about how to use one of these functions, just click on that function and a description of the function will appear on the right:

We can combine functions of different types grouped under **Aggregates**, **Arrays**, **Color**, **Conditionals**, **Conversions**, and others, to modify and/or generate geometries according to our requirement.

Working with the Data Plotly plugin

This **Data Plotly plugin** allows you to draw JavaScript library **D3** plots in QGIS and is a new addition to QGIS. You can use this plugin by installing the **Data Plotly** plugin from the **Plugins** menu in the menu bar. For this, we will use the shapefile `BGD_adm3_data_re.shp`. Now, to use this plugin, click **Plugins | Data Plotly | DataPlotly**. This opens a panel on the right-hand side. We will now make a box plot for the **value2** field of this layer. To do so, select **BGD_adm3_data_re** for **Layer** and select `value2` for **Y Field**. Note that there is also the **Grouping Field** option, which allows us to get multiple bar plots grouped by the field selected here. We can also change the color properties of this plot along with the transparency. This panel will look like this:

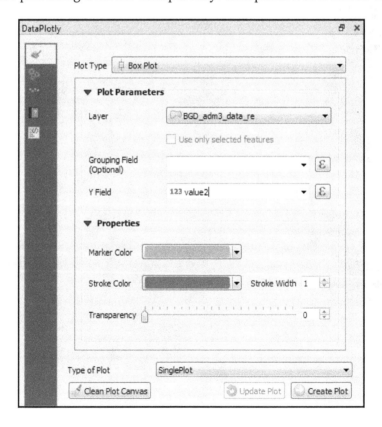

Now, click on **Create Plot** at the bottom right to get a box plot of the field `value2`. This will give the following box plot. If you hover the mouse over this boxplot, you will see the minimum, maximum, median, and the first and third quartile values:

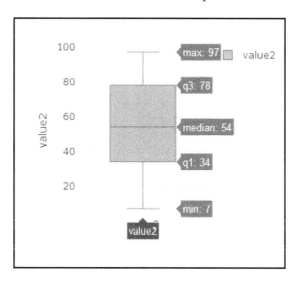

Now, to modify the existing plot or to generate a new plot, click on this tab at the top-left of this panel. Now, group this box plot according to the NAME_1 field by selecting this field for **Grouping Field (Optional)** as follows:

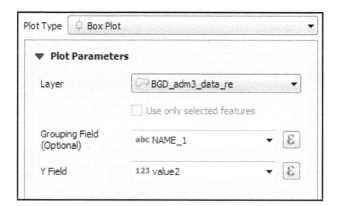

Now, click on **Update Plot** 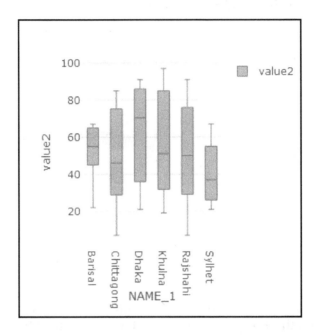; this will now show us a new box plot grouped by the NAME_1 field:

The different types of plot that you can generate can be found by clicking on **Plot Type**. You can see from the following screenshot that it offers different types of plots, such as **Bar Plot**, **Histogram**, and **Polar Plot**:

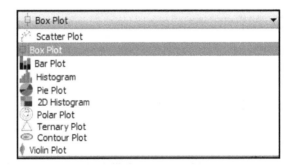

Now we can further customize the plot, such as changing the plot's title, legend title, label of the *x* axis, label of the *y* axis, by clicking on this tab, ⚙ on the left. We find the plot under this tab ▥ . We can get help with different types of plots from this tab and the ▣, and we can get the HTML code used to generate a plot from this tab ▦. After clicking this tab, you can see HTML code that you can just copy and paste for further use.

You now have a basic understanding of this plugin and so, just by exploring this with different plot types and options, you will become familiar with all the functionalities of this plugin in no time.

Summary

This chapter covered more advanced ways to style and visualize GIS data. We provided examples of using live layer effects and provided two use cases for inverted polygon shapeburst fills. We also gave examples of how to implement the new **2.5D** renderer and went on to show you how to create a 3D scene using the **QGIS2ThreeJS** plugin. Finally, we provided step-by-step instructions for creating a series of maps with the Atlas feature of Print Composer. With the information in this chapter, you will be able to generate stunning graphics.

In the next chapter, we will switch from cartography to performing spatial analyses. We will cover the **QGIS Processing toolbox**. We will begin with a comprehensive overview and a description of the layout of the toolbox. We will then explore the various algorithms and tools that are available in the toolbox, with real-world examples.

Section 4: Becoming a Master

4

This section will make you a pro in QGIS. We will be doing some amazing exercises using QGIS. We will see what a processing toolbox is and how it can open windows for even third-party algorithms and let you develop automatic graphic modelers that can enable you to create complex visualizations, such as atlases. We will also see Python with QGIS and write our first script to solve a problem.

In this section, we will cover the following chapters:

- Chapter 8, *The Processing Toolbox*
- Chapter 9, *Automating Workflows with the Graphical Modeler*
- Chapter 10, *Creating QGIS Plugins with PyQGIS and Problem Solving*
- Chapter 11, *PyQGIS Scripting*

8
The Processing Toolbox

In this chapter, we will explore the structure of the QGIS Processing Toolbox and identify which algorithm providers are available, and how to use these specialized algorithms. To accomplish these goals, we will ensure that the Toolbox is properly configured, use a variety of specialized vector and spatial algorithms from the **Geographical Resources Analysis Support System (GRASS)** and **System for Automated Geoscientific Automation (SAGA)** libraries, and perform hydrologic analyses using the SAGA library. We will cover the following topics in this chapter:

- Introducing the Processing Toolbox
- What's new in the Processing Toolbox?
- Viewing the Processing Toolbox
- Configuring the Processing Toolbox
- Using the Processing Toolbox
- Performing raster analysis with GRASS
- Performing analysis using SAGA
- Exploring hydrologic analysis with SAGA

Introducing the Processing Toolbox

The Processing Toolbox provides a one-stop-shop that helps you find the algorithms that not only belong to the QGIS tools, but also those belonging to some third party.

This initially allowed the algorithms that were QGIS natives and were a part of geospatial packages. They could be accessed only by using the software or through the command line. We can now access algorithms from the following providers using the Toolbox:

- QGIS geo-algorithms
- GDAL/OGR
- GRASS
- SAGA
- R
- Models
- Scripts

We will not make use of all the algorithm providers or explore all the available algorithms in this chapter; however, the last two entries in the list offer additional options for creating reusable graphical models and running Python scripts, which are covered in Chapter 9, *Automating Workflows with the Graphical Modeler*, and Chapter 10, *Creating QGIS Plugins with PyQGIS and Problem Solving*, respectively.

What's new in the Processing Toolbox?

The Processing Toolbox, starting from its third version (QGIS 3.x), has many new features added. These features give users many additional options for working within the QGIS environment and to complete their tasks with added ease. Some of the additional features starting with QGIS 3.x are listed here:

- **Tools reorganized and rewritten**: Many algorithms are reorganized and some are rewritten. Some previous algorithms (from QGIS version 2.x) are renamed, improved (**Reverse line direction, Extend lines, Offset lines**, and so on) and some new algorithms (**Sample raster values, Raster pixels to polygons, Drape features to z/m**, and so on) have also been added.
- **Edit in place mode:** Now, we can execute algorithms directly on the features from the layer and modify their attributes and geometries in place. This was added to Processing Toolbox for 3.4.
- **Store models inside project:** We can store processing models inside QGIS project files and, when we open a project, the saved model also gets loaded at the same time.

The preceding are some of the new features added to QGIS version 3. Please have a look at the web page to get an idea about all the new and improved features of QGIS processing, at `https://qgis.org/en/site/forusers/visualchangelog34/#feature-new-edit-in-place-mode`.

Configuring the Processing Toolbox

In this section, we will ensure that the Processing Toolbox is correctly configured to access and execute the algorithms within GRASS and SAGA. Many of the required libraries are automatically installed, but how you configure these tools will vary depending on your operating system and how you choose to install QGIS.

Find support for your installation as follows:

- To use some of the GRASS algorithms in this chapter, you might need to use QGIS Desktop 3.6.0 with GRASS 7.6.0 (for QGIS 3.6.0) or QGIS Desktop 3.4.5 with GRASS 7.6.0 (for QGIS 3.4.5). If you've used the OSGeo installer, you can use the advanced installer option to add GRASS 7 to your installation.
- Instructions for configuring most third-party algorithms on different operating systems can be found on the QGIS website at `https://docs.qgis.org/testing/en/docs/user_manual/processing/3rdParty.html`.

To begin configuring the Toolbox, we need to click on **Options** and configuration under **Processing Toolbox**, which is illustrated in the following screenshot. Note that if you are using a Linux distribution, this configuration can be found by navigating to **Processing | Options | Providers**:

To get started, you need to make sure that each of the providers that you intend to use is activated (with the installation of Desktop QGIS 3.6, GDAL, GRASS, and SAGA are automatically activated), and depending on your operating system and installation approach, you may need to specify the necessary folders to run R and LAStools. The next screenshot illustrates that each of the algorithms are activated and the necessary folders are specified:

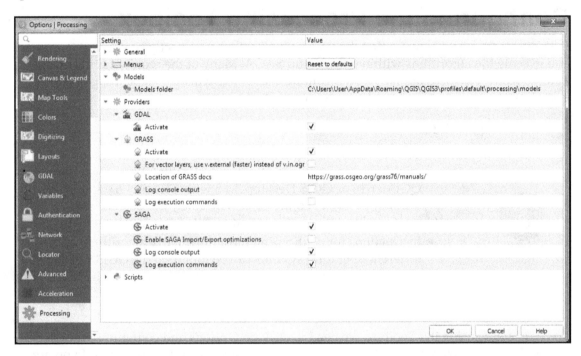

Once you click on **OK**, QGIS will update the list of algorithms accessible through the Processing Toolbox.

Viewing the Processing Toolbox

In this section, we will explore the organization of the Toolbox and establish a common language for describing its various components. Until this point, we haven't actually seen the interface itself. We've merely configured and possibly installed the required dependencies to make the **Toolbox** function. To view the Toolbox, you need to click on **Toolbox** under the **Processing** menu, as illustrated in this screenshot:

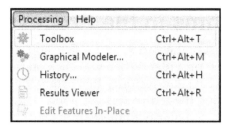

The **Processing Toolbox** will appear on the right-hand side of the QGIS interface:

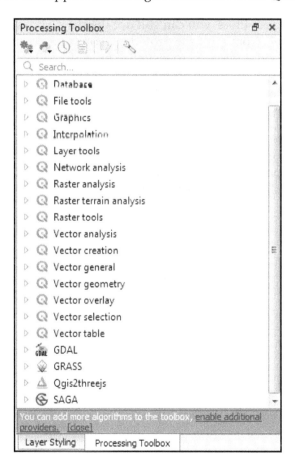

The next section will provide you with an overview of how to run an algorithm in this Toolbox.

Running algorithms in the Processing Toolbox

Initially, you will only see a list of the various providers and a summary of the total algorithms available from each provider. When you click on the ▷ icon next to any of the entries, the list expands to reveal subdirectories that group together related tools. In addition to manually navigating these directories, there is a search box at the top of the Toolbox. So, if you are already familiar with these third-party packages or are looking for a specific tool, this may be a more efficient way to access the algorithms of interest.

You can search algorithms by topic: Even if you aren't familiar with the algorithm providers, you can still use the search box to explore what tools are available from multiple providers. For example, if you are interested in finding different ways to visualize or explore topographic relationships, you could search for it by typing `topographic` in the textbox and discover that there are 9 tools from three different providers that relate to topography!

To open any algorithm of interest, you just need to double-click on the name and the algorithm dialog interface will open. It looks similar to other tools that we have already used in QGIS. For example, click on the ▷ icon next to the **GDAL** entry, then **Raster analysis**, and double-click on **Aspect** as shown here:

You will see the interface shown in the following screenshot:

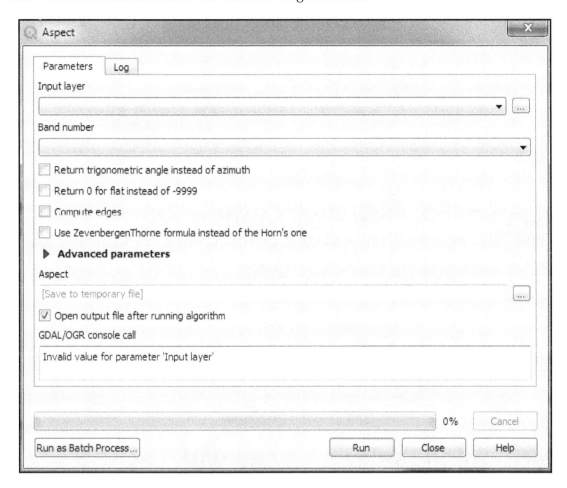

Any algorithm that you select will present you with a similar algorithm dialog box, so it is worth exploring the functionality of the interface. Similar to the other tools that we have already used, clicking on any play sign will reveal drop-down options that will allow us to select an option that is passed to the algorithm. The dialog boxes within the Processing Toolbox also provide a tab that may provide additional information. The **Log** tab will record the history of any operations performed using the tool, which is often useful for debugging errors.

You can explore the functionality of each of the algorithms that we are going to use in this section by going to the official website of each provider:
GRASS: https://grass.osgeo.org/
SAGA: http://www.saga-gis.org/en/index.html

Now that we have a basic understanding of its overall functionality and organization, we can begin using it to utilize tools that weren't historically available within the QGIS environment. The intent with these exercises isn't to provide a comprehensive overview of all the providers or algorithms, but to illustrate the power and flexibility that the Toolbox brings to QGIS.

Using the Processing Toolbox

We will begin by using some of the GRASS algorithms and focusing primarily on tools that aren't available through default plugins or drop-down menus. For example, even though GRASS has the ability to calculate aspect, this functionality is already available in QGIS and can be found by navigating to **Raster | Analysis**.

The original data used in this chapter can be obtained from these sources:

- http://oe.oregonexplorer.info/craterlake//data.html
- http://www.mrlc.gov/
- http://www.opentopography.org/

Performing raster analysis with GRASS

The **Geographical Resources Analysis Support System** (**GRASS**) environment represents one of the first open source GIS options available. It has a long history of providing powerful geospatial tools that were often overlooked because the GRASS interface and data organization requirements weren't as intuitive as other, often proprietary, options. The integration of GRASS algorithms within the Processing Toolbox provides access to these powerful tools within an intuitive GUI-based interface.

To explore the types of GRASS algorithms available through the Toolbox, we will work through a series of hypothetical situations and perform the following analyses:

- Calculating a least-cost path across a landscape
- Evaluating a viewshed

Please make sure that you have downloaded, unzipped, and added the necessary data to QGIS and set the project CRS value to EPSG: 26710. We need to organize this data so that the elevation layer is at the bottom of the data layer panel, as illustrated in the next screenshot:

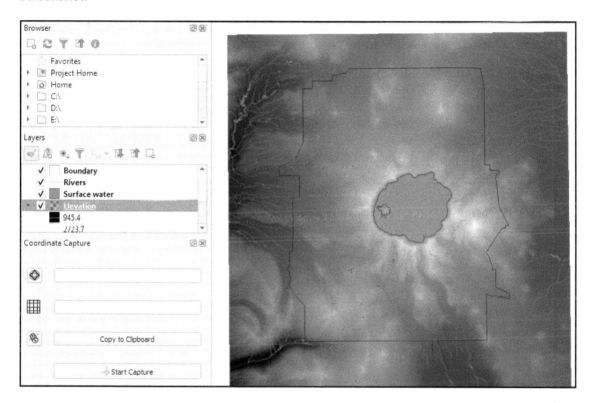

The ZIP folder contains the following files:

- An elevation file (dems_10m.dem)
- A boundary file (crlabndyp.shp)
- A surface water file (hydp.shp)
- A land use file (lulc_clnp.tif)
- A search and rescue office file (Start.shp)
- An injured hiker file (End.shp)
- A fire towers file (towers.shp)

Calculating shaded relief

The basic requirement for many of the tools within the GRASS library is a **Digital Elevation Model** (DEM) or **Digital Terrain Model** (DTM). However, since a DEM is a layer that contains continuous data representing elevation, when we load a DEM into QGIS, or any GIS for that matter, it has a flat appearance. Therefore, it is sometimes difficult to visually evaluate how topography might influence the results of our analyses.

So, our first foray into the GRASS library will make use of the **r.shaded.relief** tool to create a shaded relief map or **hillshade**, which can provide some topographic context for spatial analyses. Remember that you can access this algorithm by using the **processing commander**, the **searchbar**, or by navigating through the GRASS GIS commands list. Once the dialog box is open, we need to select the elevation layer of interest (in this case, the elevation layer). Leave all the default parameters the way they are, and click on **Run**:

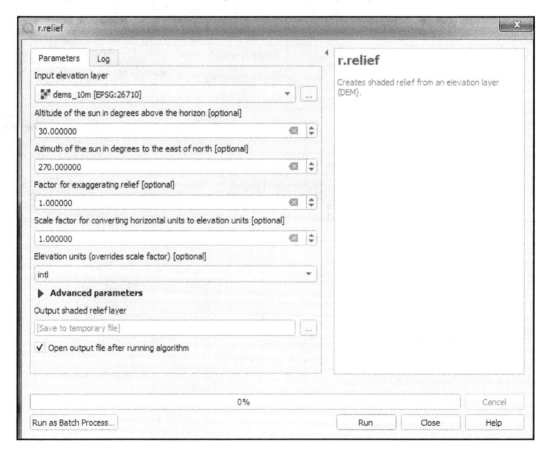

Changing the default save option

By default, QGIS saves any new layer as a temporary file in memory. To save all your output files to a directory, you need to click on the small square button containing three dots and specify the location where the file needs to be saved.

We could have easily used the built-in **Terrain Analyses** tools and executed the **Hillshade** tool by navigating to **Raster | Terrain Analyses**, but the decision to use GRASS was deliberate to illustrate that, more often than not, algorithms in the Processing Toolbox offer more optional parameters for better control over the resulting output. In this case, the output can be moved to the bottom of the data layer panel and the blending mode of the elevation layer can be set to Darken. The results of this blending operation are shown in the following screenshot:

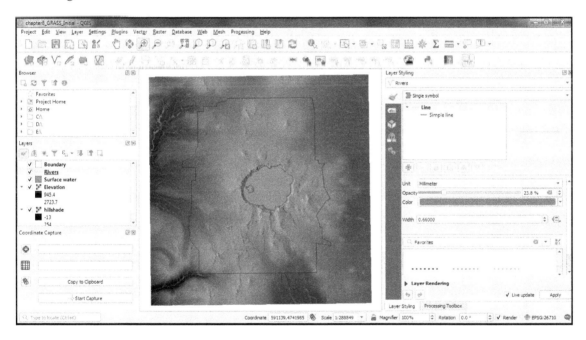

Although this tool would typically be considered more of a geoprocessing action than an analytical tool, this type of algorithm has been used to evaluate topographic shading at various times throughout a given year to estimate the persistence of snow and characterize potential habitat. If the intent is to merely show a visualization and not perform any spatial analyses, remember (from Chapter 2, *Styling Raster and Vector Data*) that we can symbolize the elevation layer using colors rather than grayscale to better visualize changes in elevation.

Adjusting default algorithm settings for cartographic reasons:
The default altitude and azimuth settings specify the position of the sun relative to the landscape, which isn't an unrealistic value for Crater Lake. However, it is possible to move the sun to an unrealistic position to achieve better contrast between topographic features.
For an extensive exploration of shaded relief techniques, visit http://www.shadedrelief.com/.
To calculate accurate azimuth and elevation values for varying latitudes, visit http://www.esrl.noaa.gov/gmd/grad/solcalc/azel.html.

Calculating least-cost path

Least-cost path (LCP) analyses have been used to model historical trade routes and wildlife migration corridors, plan recreation and transportation networks, and maximize safe backcountry travel in avalanche-prone areas, to name just a few applications. To perform an LCP analysis in QGIS, we are going to use a variety of tools from the Processing Toolbox and combine the resulting output from the tools.

Although there are numerous useful geoprocessing algorithms in the GRASS library, we are going to focus on more advanced spatial analyses that better demonstrate the analytical power residing in the Processing Toolbox. We are going to calculate a least-cost path for a hypothetical situation where a search and rescue team has been deployed to Crater Lake National Park to extract an injured hiker. The team may be able to use roads for part of their approach, but would like to identify the least cost or the least rigorous approach to the hiker. Essentially, we are going to make some simplistic assumptions about how much effort will be required to move across the landscape by incorporating slope and land use into a raster layer representing the cost of movement. This cost layer will then be used to identify the least-cost path from the search-and-rescue office to the injured hiker.

In order to accomplish this analysis, we need to complete the following steps:

1. Calculate the slope using r.slope.aspect
2. Reclassify the new slope raster using the rules in the _recode.txt slope file
3. Reclassify the land use raster using the rules in lulc_reclass.txt
4. Combine the reclassified slope and land use layers.
5. Calculate the cumulative cost raster using r.cost
6. Calculate the cost path using LCP

Calculating slope using r.slope

We will now calculate slope using a digital elevation model file and use `r.slope.aspect` from **GRASS**. The necessary settings for calculating slope are illustrated in the following screenshot:

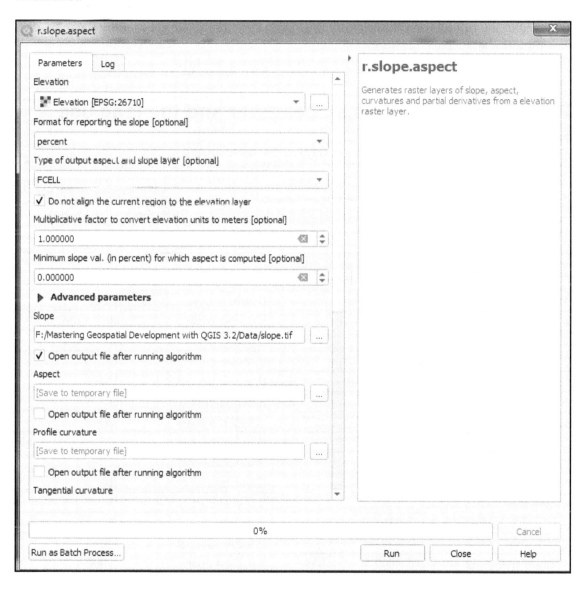

Using **QGIS Desktop 3.6** (or **QGIS Desktop 3.4**), if you see error in opening this tool from GRASS, then, close **QGIS Desktop 3.6.0 (or 3.4)** and rather open **QGIS Desktop 3.6.0 with GRASS 7.6.0 (or QGIS Desktop 3.4.5 with GRASS 7.6.0)**.

This dialog box indicates that slope is being calculated in percent, which will need to be reflected in the ruleset used to reclassify this layer. Essentially, we are making the assumption that increasing slope values equate to increased physical exertion and thus inflict a higher cost on members of the search-and-rescue team.

Reclassifying the new slope raster and the land use raster

To accomplish this, we are going to input the slope raster into `r.recode` and use the slope `_recode.txt` file to inform the tool how to reclassify the slope values. It is worth opening up the slope `_recode.txt` file to understand the GRASS formatting requirements and evaluate the assumptions within this re-classification scheme, which are also summarized in the following table:

Land use/Land class type	Land use/Land class code	Travel cost assumption	Recode value
Water	11	Highest cost	1000
Developed open land	21	Lowest cost	1
Developed low intensity	22	Lowest cost	1
Developed medium intensity	23	Lowest cost	5
Developed high intensity	24	Moderate cost	20
Barren land	31	Lowest cost	1
Deciduous forest	41	Moderate cost	20
Evergreen forest	42	Moderate cost	50
Mixed forest	43	Moderate cost	20
Shrub/scrub	52	Low cost	10
Grassland	71	Lowest cost	5
Pasture/hay	81	Lowest cost	5
Cultivated crops	82	Moderate cost	20
Woody wetlands	90	Highest cost	1000
Wetlands	95	Highest cost	1000

The following screenshot illustrates how to populate the `r.recode` algorithm:

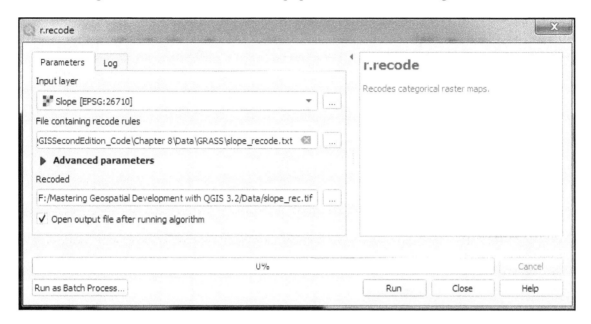

We need to use this same tool to recode the values of the provided land use layer using the `lulc_recode.txt` ruleset. Again, it is worth exploring this file to evaluate the assumptions made about the costs for moving through each land use classification; for example, we have assumed that water has the highest cost and developed open space has the lowest cost. To properly explore this layer, you will need to import the `lulc_palette.qml` QGIS style file, which will categorize land use by name (for example, water, mixed forest, and so on).

Combining reclassified slope and land use layers

Once we have created both the slope and land use cost grids, we can combine them using the native QGIS **Raster calculator** tool to use with the **r.cost** algorithm. Since neither layer contains any zero values, which would need to be preserved through multiplication, we can combine them using addition:

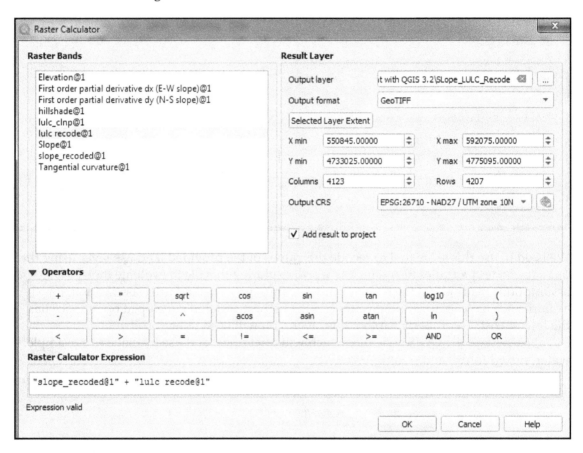

We could also use the **r.mapcalculator** tool to combine these layers, but this demonstrates how easy it is to move between native QGIS tools and those housed in the Processing Toolbox.

In this example, we used **r.cost** to create a new layer representing the cost of traveling across the landscape. If we know that the path will be traveled exclusively on foot, it may make more sense to use the **r.walk** algorithm available through the GRASS library. For more information about this, visit `https://grass.osgeo.org/grass76/manuals/r.cost.html`.

Calculating the cumulative cost raster using r.cost

To summarize our progress so far, we have reclassified the slope and land use layers and combined them so that we can now create a cost grid that can be used to evaluate the cost associated with moving between individual cells. This analysis requires two additional inputs, a starting and ending point that are provided as separate shapefiles, `Start.shp` and `End.shp`. These points serve as a guide for how the algorithm should characterize the cost of moving through an area of interest. The following screenshot illustrates how to populate the **r.cost** tool:

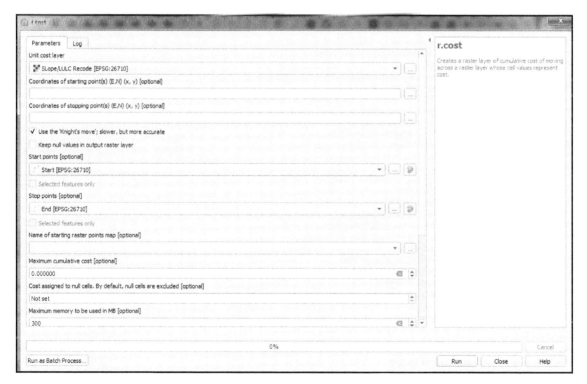

Calculating the cost path using LCP

This land use layer can now be used to calculate the LCP between each individual cell. To accomplish this, we are going to make use of a tool from the SAGA library, which is explored in more depth later in this chapter. This approach again demonstrates the flexibility of the Processing Toolbox and how easy it is to combine tools from various libraries to perform spatial analyses. We need to search the Toolbox for **LCP** and identify the relevant point (in this example, there will be only two point layers, so SAGA finds them by default), specify the cost grid, and then define an output for the resulting LCP, as illustrated in the following screenshot:

Now, we just need to organize the relevant layers, as shown in the next screenshot, to inform the Crater Lake search-and-rescue team about the least-cost approach to the injured hiker:

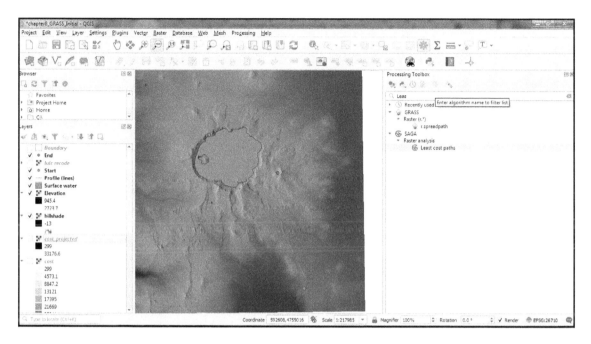

In this exercise, we made use of both GRASS and SAGA algorithms to calculate an LCP. These algorithms allowed us to calculate a hillshade and slope, reclassify raster layers, combine raster layers, create a cost grid, and calculate an LCP from this cost grid. Although this exercise was clearly hypothetical and limited in the number of parameters used to evaluate cumulative cost, it hopefully demonstrates how easy it is to perform this type of analysis. LCP analyses have been used to model historical trade routes *(Howey, 2007)*, wildlife migration corridors *(Morato et al. 2014)*, plan recreation and transportation networks *(Gurrutxaga and Saura, 2014)*, and maximize safe backcountry travel in avalanche-prone areas *(Balstrøm, 2002)*, to name just a few applications.

For more information, please see the following references:

- *Howey, M. (2007). Using multi-criteria cost surface analysis to explore past regional landscapes: a case study of ritual activity and social interaction in Michigan, AD 1200-1600. Journal of Archaeological Science, 34(11): 1830-1846*
- *Morato, R. G., Ferraz, K. B., de Paula, R. C., & Campos, C. d. (2014). Identification of Priority Conservation Areas and Potential Corridors for Jaguars in the Caatinga Biome, Brazil. Plos ONE, 9(4), 1-11. doi:10.1371/journal.pone.0092950*
- *Gurrutxaga, M. and Saura, S. (2014). Prioritizing highway defragmentation locations for restoring landscape connectivity. Environmental Conservation, 41(2), 157-164. doi:10.1017/S0376892913000325*
- *Balstrøm, T. (2002). On identifying the most time-saving walking route in a trackless mountainous terrain, 102(1), 51-58. 10.1080/00167223.2002.10649465*

Evaluating a viewshed

Another advanced spatial analysis technique involves evaluating viewsheds to address the intervisibility between features, or the potential visual impact of vertical structures, such as wind turbines or radio and cell towers. This type of analysis is often incorporated into an environmental impact evaluation, but the technique has other applications, such as evaluating which proposed viewing platform offers the greatest viewable area, or determining how best to position observers during an aerial threat assessment. Although this tool has a specific niche application, working through this section will allow us to make use of additional algorithms that have broader applications.

We will begin by creating a new QGIS project and adding the following files:

- The elevation file (dems_10m.dem)
- The boundary file (crlabndyp.shp)
- The surface water file (hydp.shp)
- The fire towers file (towers.shp)

In this application, we are going to assume that the **National Park Service** (**NPS**) has asked us to evaluate the visual impact of building three proposed fire towers. We need to perform a viewshed analysis and provide an estimate of the total area impacted within the park.

In order to accomplish this analysis, we need to complete the following steps:

1. Clip the elevation to the boundary of the park using GDAL
2. Calculate viewsheds for towers using **r.viewshed**
3. Combine the viewsheds using **r.mapcalculator**

Clipping elevation to the boundary of the park using GDAL

To reinforce the concept that we can make use of a variety of algorithms within the Processing Toolbox to accomplish our analyses, we will use the **Clip Raster by Mask Layer** tool that is available through the GDAL/ORG algorithms. We will clip the elevation layer to the park boundary so that we save processing time by only evaluating the viewshed within the park. We can find this tool by typing `clip` in the search bar. The following screenshot illustrates how to set the parameters for this tool:

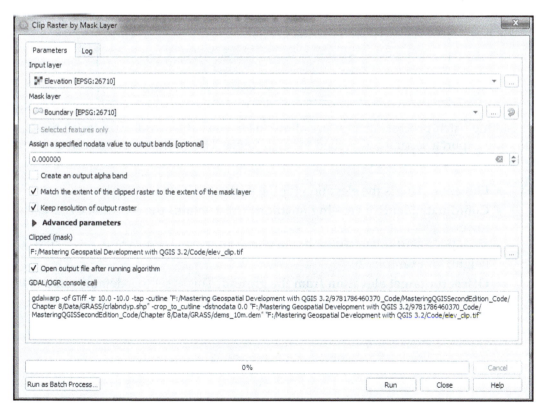

Calculating viewsheds for towers using r.viewshed

Once we have a clipped elevation layer, we can set the transparency of the value *0* to *100* percent in the transparency tab of the **Layer Properties** interface and begin the process of calculating the viewshed using the **r.viewshed** tool. If you open this tool using the processing commander or double-click on the entry within the Toolbox, you will be presented with a dialog box that contains the option to enter a coordinate identifying the viewing position. As we have three towers of interest, we will manually execute this tool three different times. However, most of the algorithms in the Toolbox have the option to **Execute as Batch Process.** By right-clicking on the tool, we can select this option, as illustrated in the following screenshot:

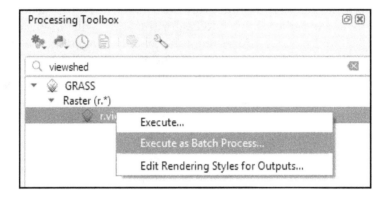

The resulting batch-processing interface allows us to enter the coordinates for all three towers and export a separate viewshed for each tower. The Toolbox options can be set to the following values:

- **Elevation**: This is the elevation for the park.
- **Coordinate identify viewing position**: This contains the coordinates for each tower.
- **Viewing position elevation above the ground**: This is the viewing position height; for example, 20.
- **Offset for target elevation from the ground**: This is set to a default value of 0.
- **Maximum visibility radius. By default infinity (-1)**: This option is set to a default value of -1 or infinity. We provide a value of 32000.

- **Refraction coefficient**: The refractivity coefficient of light.
- **Amount of memory to use in MB**: This option is set to 500.
- **Intervisibility**: The default is set to blank. Click on **...** to save as a separate file.

If we had more than three towers, we could click on the **Add row** button at the bottom of the batch-processing interface.

We can begin entering the necessary parameters using the coordinates provided in the following table and the guidelines:

Tower number	Coordinates
1	574599.082827, 4749967.314004
2	580207.025953, 4752197.343687
3	571656.050455, 4750321.28697

It is worth exploring the rationale behind some of the input parameters. The first two are hopefully obvious—we need an elevation layer and observer points to evaluate the viewshed for any assumptions. However, setting the position height above ground to 20 meters is an average value for typical fire towers. The maximum distance of 32,000 meters is the greatest distance between any of the towers and the edge of the park elevation layer, and including Earth's curvature – even for small areas – at worst increases processing time, but provides a more accurate representation of visibility.

If you have a lot of observers, completely fill out the information for the first observer, and after you set the **Output raster layer** parameter, you will be prompted to autofill the input boxes. If you select `yes`, the interface will automatically populate the parameters and you will only need to adjust the parameters that are different. For example, the coordinates will need to be updated, and perhaps not all observers have the same height.

We will manually use **r.viewshed** three different times, each time with different coordinates given in the table. The following is a snapshot of the first time we use **r.viewshed** with the first set of coordinates:

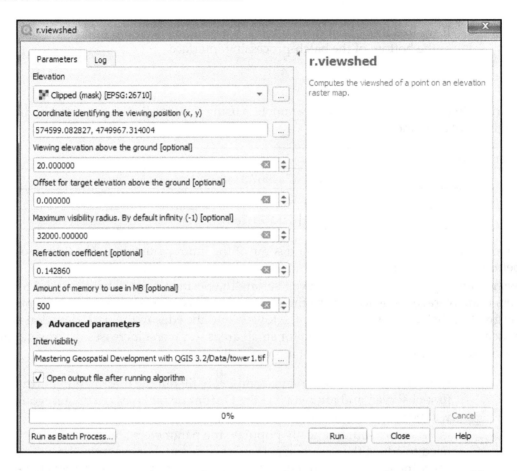

We need to repeat this step for the other two coordinates given in the previous table, and save them as **tower 2.tif** and **tower 3.tif**.

Combining viewsheds using r.mapcalc.simple

In order to evaluate the cumulative visual impact of all the three towers, we need to add them together. However, the algorithm outputs a grid that contains either a degree angle representing the vertical angle with respect to the observer, or null values.

If we attempt to add three layers that contain null values, the resulting output will not accurately reflect the total visible area within the park. To address this issue, we need to make use of the `isnull` function within **r.mapcalc.simple**. We will use this function within a conditional statement to identify where there are null values and replace them with a zero so that we can accurately combine all the three layers. We need to open **r.mapcalc.simple** and write this conditional statement:

```
if(isnull(A),0,1)+ if(isnull(B),0,10) + if(isnull(C),0,100)
```

The query that we are asking the calculator to execute is as follows: if layer A is null, then replace it with a value of zero, otherwise give the resulting grid a value of 1 and then add it to the results from the other three layers, which are also evaluated for null values. By replacing the original values with any one of 0, 1, 10, or 100, we are able to evaluate the total cumulative viewshed and also differentiate between the impacts of individual towers. The following screenshot illustrates how to ask these questions within the raster calculator:

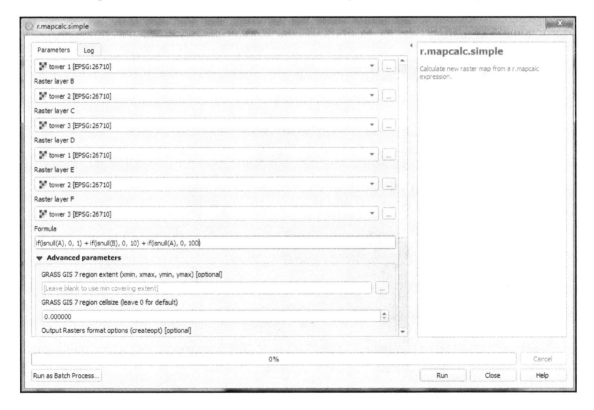

The resulting output will contain values that can be used to interpret which towers contribute to the cumulative viewshed. These values are summarized in the following screenshot. To better visualize the cumulative viewshed within the park, you can load the view_style.qml layer and adjust the colors to your preference, as follows:

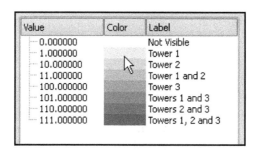

We can also provide a more informative visual depiction of the impact, as demonstrated in the next screenshot, using this approach, rather than the traditional binary visible/not-visible viewshed maps:

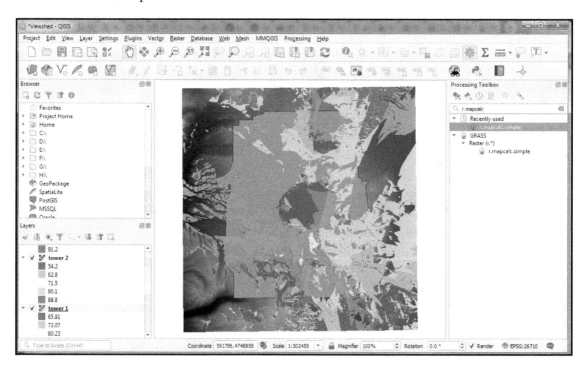

In this exercise, we used a variety of GRASS algorithms to explore the analytical power of the Processing Toolbox. We performed both common geoprocessing and advanced spatial analyses to arrive at hypothetical scenarios that would be time-consuming to address without the support of a GIS; these analyses included the following:

- Creating a shaded relief map using **r.shaded.relief**
- Calculating slope using **r.slope**
- Reclassifying raster data using **r.recode**
- Creating a cost grid using **r.cost**
- Calculating an LCP using LCPs
- Calculating a viewshed using **r.viewshed**
- Utilizing raster calculation functions within **r.mapcalc.simple**

In the next section, we will continue exploring the types of analyses possible using the SAGA algorithms that are available through the Toolbox.

Performing analysis using SAGA

The **System for Automated Geoscientific Automation** (**SAGA**) environment contains powerful tools, some of which have very specific applications; for example, geostatistical analyses or fire and erosion modeling. However, we will explore some of the SAGA tools that have broader applications and often dovetail nicely with tools from other providers. Similar to GRASS, integrating the SAGA algorithms within the Processing Toolbox provides access to powerful tools within a single interface.

To explore some of the SAGA algorithms available through the Toolbox, we will work through a hypothetical situation and perform the analysis to evaluate the potential roosting habitat for the northern spotted owl.

We are going to continue using data from the provided ZIP file, and we will need the following files:

- The elevation file (dems_10m.dem, available in the GRASS data folder)
- The hillshade file (hillshade.tif, created in the GRASS section)
- The boundary file (crlabndyp.shp)
- The surface water file (hydp.shp)
- The land use file (lulc_clnp.tif, available in the GRASS data folder)

Evaluating a habitat

GIS has been used to evaluate potential habitats for a variety of flora and fauna in diverse geographic locations. Most of the habitats are more sophisticated than the approach we will take in this exercise, but the intention is to demonstrate the available tools as succinctly as possible. However, for simplicity's sake, we are going to assume that the resource management office of Crater Lake National Park has requested an analysis of a potential habitat for the endangered northern spotted owl. We are informed that the owls prefer to roost at higher elevations (approximately 1,800 meters and above) in dense forest cover, and in close proximity to surface water (approximately 1,000 meters).

In order to accomplish this analysis, we need to complete the following steps:

1. Calculate elevation ranges using the SAGA **Raster calculator** tool
2. Clip land use to the park boundary using **Clip grid** with **polygon**
3. Query land use for only surface water using the SAGA **Raster calculator**
4. Find the proximity to surface water using GDAL **Proximity**
5. Query the proximity for 1,000 meters of water using the GDAL **Raster calculator**
6. Reclassify land use using the **Reclassify grid values** tool
7. Combine raster layers using the SAGA **Raster calculator**

Calculating elevation ranges using the SAGA Raster calculator

There are multiple ways to create a layer that represents elevation ranges or, in this case, elevation zones that relate to potential habitats. One method would be to use **r.recode** as we did in the GRASS exercise; another would be to use the **Reclassify Grid Values** tool provided by SAGA, which we will use later in this exercise; but another very quick way is to only identify the areas above a certain elevation – in this case, greater than 1,800 meters – using a raster calculator. This type of query will produce a layer with a binary level of measurement, meaning the query is either true or false. To execute the raster calculator, select the layer representing elevation only in the park, enter the gt(a, 1800) formula, name the output file, and click on **OK**.

The syntax we entered in the formula box tells the SAGA algorithm to look at the first grid - in this case, a - and if it has a value greater than (gt) 1,800 meters, the new grid value should be one, otherwise it should be zero. The following screenshot illustrates how this appears in the SAGA **Raster calculator** window. We could have also used the native QGIS **Raster calculator** tool. So, the intent here is to demonstrate that there are numerous tools at our disposal in QGIS that often perform similar functions. However, the syntax is slightly different between the QGIS, GRASS, and SAGA raster calculators, so it is important to check the **Help** tab before executing each of the tools:

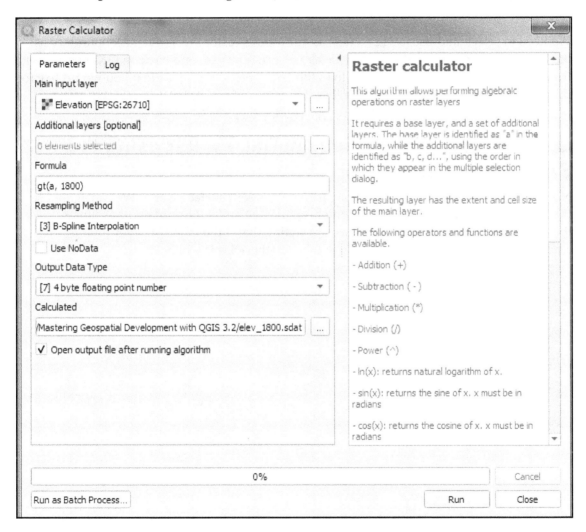

Clipping land use to the park boundary using Clip grid with polygon

After executing this tool, we will be presented with a new raster layer that identifies the elevations above 1,800 meters with a value of 1 and all other values with a value of 0. The next step is to clip the land use layer using the SAGA **Clip raster by mask layer** tool. If you remember, we clipped a raster layer in a previous exercise using the native **GDAL Clipper** tool, so again, this is merely demonstrating the number of options we have to perform spatial operations.

We need to select land use (`lulc_clnp`) as our input's raster layer, the park boundary as our polygon's layer, name the output file as `lulc_clip.tif`, and click on **OK**. Remember, from an earlier exercise, that you can load the `lulc_palette.qml` file if you would like to properly symbolize the land use layer, but this step isn't necessary.

Querying land use for only surface water using the SAGA Raster calculator

Now, we can query this layer for the areas that represent surface water. We can again use the SAGA **Raster calculator** tool and enter `eq(a, 11)` in the **Formula** box, as illustrated in the next screenshot. In this example, we are stating that if the land use layer (that is, a) is equal to 11, the resulting output value will be 1; otherwise it will be 0:

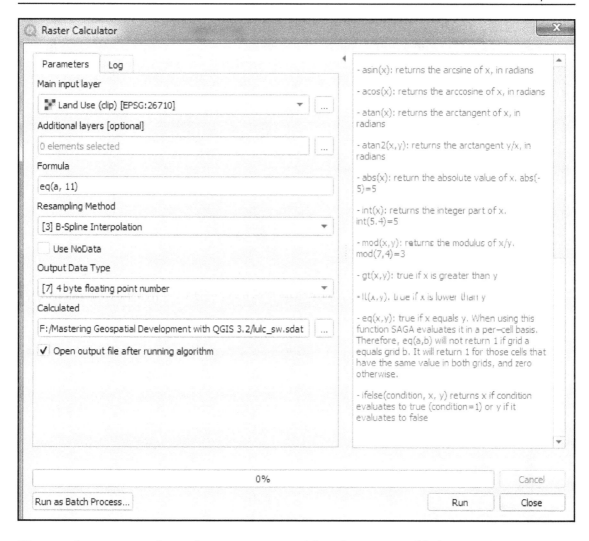

Now, we have a raster layer that we can use to identify a potential habitat within 1,000 meters of surface water.

Finding proximity to surface water using GDAL Proximity

We need to create a layer representing proximity to surface water and query that layer for areas within 1,000 meters of any surface water. Our first step is to execute the GDAL **Proximity (raster distance)** tool in the Processing Toolbox.

We need to select the binary (true or false) layer representing surface water (`lulc_sw.tif`), set the **A list of pixel values in the source image to be considered target pixels** field to 1, change the **Distance units** field to **Georeferenced coordinates**, change the **Output raster type** to **Int32**, leave all the other defaults as they are, and name the output layer, as illustrated in the next screenshot:

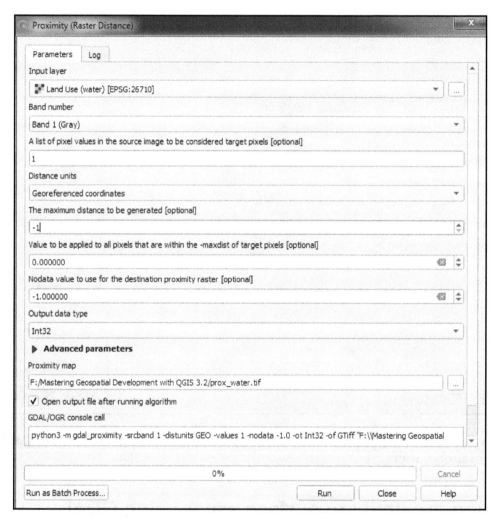

The rationale for setting **A list of pixel values in the source image to be considered target pixels** to 1 and **Distance units** to **Georeferenced coordinates** is that we are asking the algorithm to assume that the distance is measured in increments of 1, based on the geographic distance – in this case, meters – and not on the number of pixels.

We can now query the resulting grid for the area that is within 1,000 meters of surface water, but it is important to recognize that we want to identify the areas that are less than 1,000 meters of surface water, but greater than 0. If we just query for values less than 1,000 meters, we will produce an output that will suggest that the roosting habitat exists within bodies of water.

Querying the proximity for 1,000 meters of water using the GDAL Raster calculator

The easiest way to perform this query is by using the native QGIS **Raster calculator** tool by clicking on **Raster Calculator** under **Raster**. The following screenshot illustrates how to enter the "Proximity to Water@1" > 0 AND "Proximity to Water@1" <= 1000 syntax to identify a range between 0 and 1,000 meters:

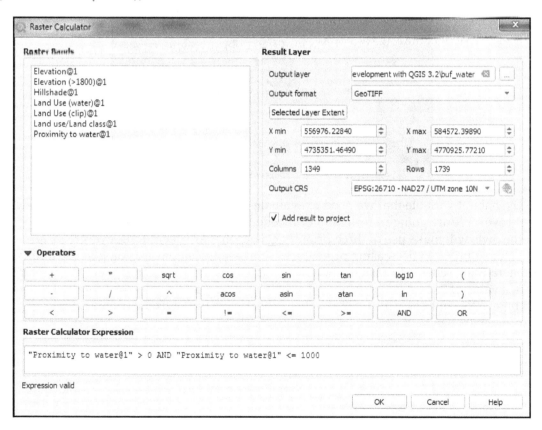

The resulting output will contain the values of 0 and 1, where 1 represents the cells that are within 1,000 meters of surface water and 0 represents those that are beyond 1,000 meters. The next screenshot illustrates what this layer looks like after you set the value of 0 to transparent, and the buffer itself to red:

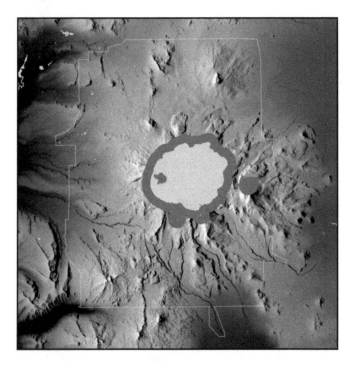

The last habitat variable that we need to evaluate is the preference toward roosting in dense forest cover. We are going to reclassify or recode the land use layer with the assumption that the owls will make use of three primary classes that are deciduous, evergreen, and mixed forest types, in decreasing preference. This means that we are going to use a simple ordinal ranking scheme to assign a new value of 3 to grid cells that represent evergreen forest, a value of 2 for shrub cover, a value of 0 for water, and a value of 1 for the remaining land types. We could assume 0 for all other cover types, but owls are unpredictable.

Reclassifying land use using the Reclassify grid values tool

To accomplish this, we could use **r.recode**, as we did in a previous exercise, but instead, we are going to use the SAGA **Reclassify Values (Simple)** tool:

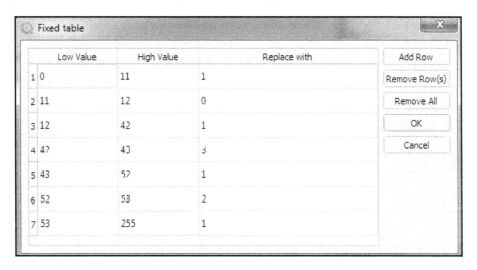

	Low Value	High Value	Replace with	
1	0	11	1	
2	11	12	0	
3	12	42	1	
4	42	43	3	
5	43	52	1	
6	52	53	2	
7	53	255	1	

If your version of QGIS doesn't allow you to add or remove rows, remember that you can also use the GRASS **r.recode** tool after creating a recode rule file. This might be a good exercise to work through to make sure you understand the formatting requirements for GRASS recode rule files. For a more in-depth explanation, visit `https://grass.osgeo.org/grass76/manuals/r.recode.html`.

To use the SAGA **Reclassify Values (Simple)** tool, we need to provide an input grid, which in this case is the clipped land use layer (`lulc_clip.tif`), and set **Method** to `[2] Low value <= grid value < high value`. The values that need to be reclassified (as shown in the previous screenshot) can be entered by clicking on the **Fixed table 0x3** button located below the **Lookup Table** prompt. Make sure you provide a name for the new reclassified grid, for example, `lulc_rec`.

Combining raster layers using the SAGA Raster calculator

Now, we have all the necessary layers to finalize our simplistic model of the northern spotted owl habitat. Since we have zero values that need to be preserved, that is, places where owls will never roost, we will multiply the three layers together using the QGIS **Raster calculator** tool. The next screenshot illustrates how to populate the raster calculator by selecting the reclassified elevation layer (`elev_1800.tif`) as the main input layer, and then selecting the reclassified water-proximity (`buf_water.tif`) and land use (`lulc_rec.tif`) layers by clicking on the ellipsis button (...) next to **Additional layers [optional]**:

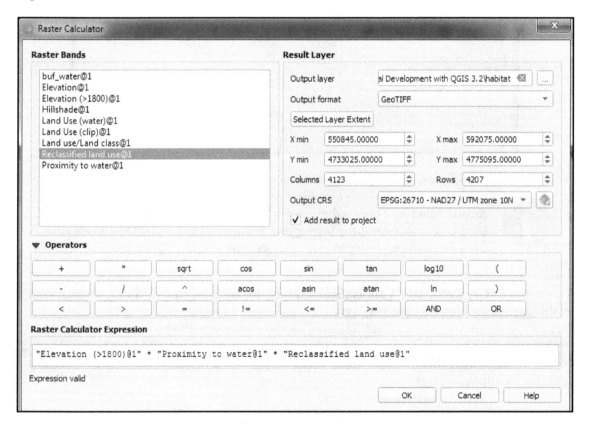

We could have used the SAGA **Raster calculator** or GRASS **r.mapcalculator** tool, but once again, this demonstrates how easy it is to switch between the various Toolbox options. To ensure that we understand the values reported in the resulting output from this calculation, we need to remember that the reclassified water and elevation layers are binary, so it will have the values of 0 and 1, while the reclassified land use layer contains the values from 0 to 3. Therefore, the new layer can only contain values of from 0 to 3, where 0 indicates no habitat, 1 indicates poor habitat potential, 2 indicates moderate habitat potential, and 3 indicates good habitat potential, as illustrated in the next screenshot:

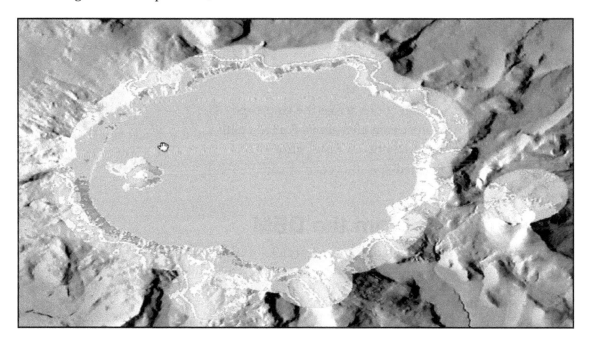

Hopefully, it is clear that this is a very simple model with many assumptions that any ornithologist who actually studies the northern spotted owl would not actually use to evaluate habitat. However, the various tools and general approach that has been taken to evaluate this hypothetical scenario could be applied by paying more rigorous attention to the underlying assumptions about the variables that influence potential habitat. The goals of working through this type of analysis were threefold: to showcase a variety of useful SAGA algorithms, to demonstrate that there are similar tools with subtle differences that are available through the Toolbox, and to illustrate how easy it is to switch between native QGIS tools and those found in the Toolbox.

Exploring hydrologic analysis with SAGA

We are going to continue using data from the provided ZIP file, and we will need the following files:

- Elevation file (`dems_10m.dem`, available in the GRASS data folder)
- Gauge shapefile (`gauge.shp`)
- Rivers file (`hydl.shp`)

To explore the hydrologic functionality of SAGA, we will characterize the watershed of Sun Creek upstream to the town of Fort Klamath, California. To accomplish this, we will perform the following tasks:

1. Remove pits from the DEM
2. Calculate flow directions across the landscape
3. Calculate the upstream area above Fort Klamath
4. Calculate a stream network raster grid
5. Create a watershed-specific vector stream network

Removing pits from the DEM

Before using any hydrologic algorithms, regardless of the algorithm provider, we need to make sure that the DEM is hydrologically corrected. This means that we need to ensure that it behaves like the natural landscape, where surface flow moves across the land and does not get trapped in pits or depressions, in the DEM.

To accomplish this, we are going to use the **Fill Sinks (Wang & Liu)** tool from SAGA on our new DEM file, as illustrated in this screenshot:

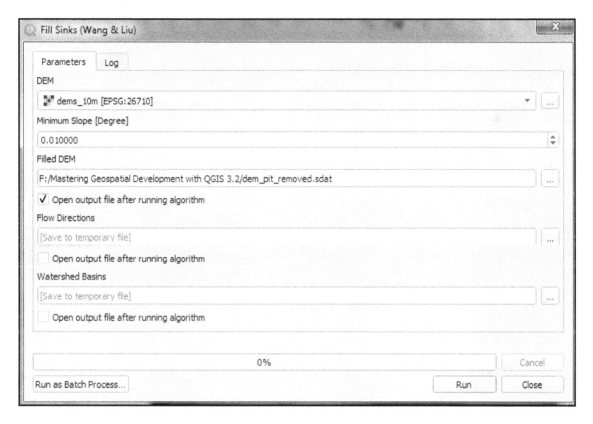

Now, go to **Layer Properties** of **Filled DEM** and, under the **Symbology** tab, go to **Color ramp | Create New Color Ramp...** and select **Catalog: cpt-city**, and click **OK**. A snapshot of these steps is shown as follows:

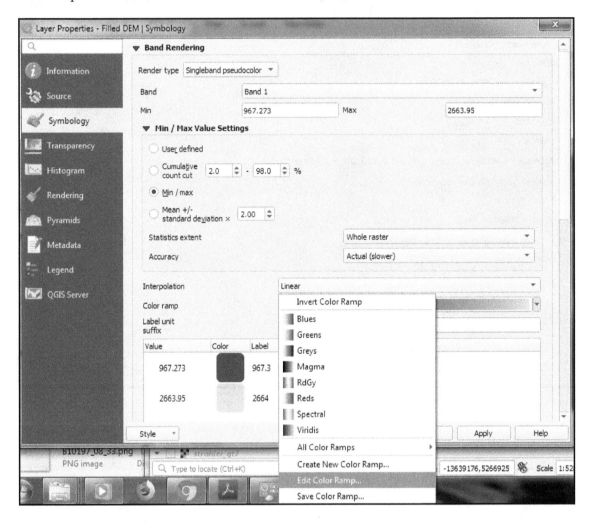

Now, we will be presented with a number of options for our color ramp. Under **Topography**, select **elevation** and click **OK**:

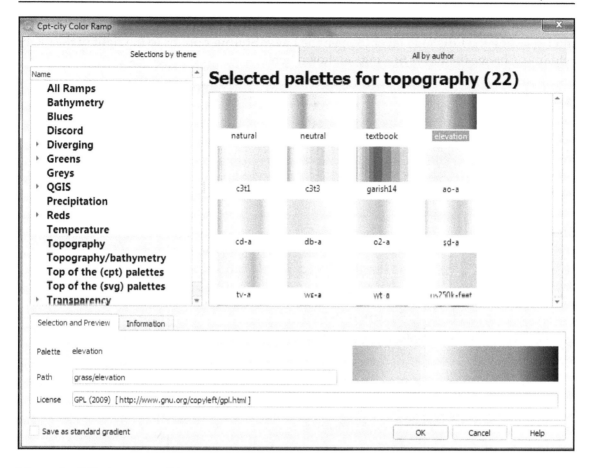

Now, the Filled DEM will have this color palette applied to it:

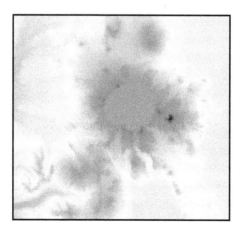

Now, copy **Filled DEM** and rename the new layer, **Filled Dem Copy**, to **hillshade**. Press **F7** after selecting the **hillshade** layer to open the **Layer Styling** options to the right of the QGIS window. Change the **Layer Styling**, as shown here:

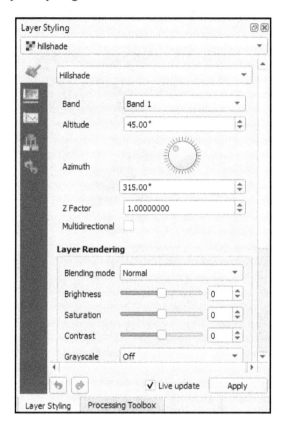

To apply the **Hillshade** style, put the **Filled DEM** above the **hillshade layer** and change the latter's **Blending mode** to **Multiply**:

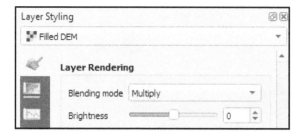

Now, the DEM will look like this:

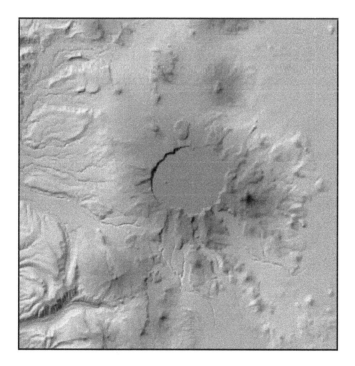

Deriving streams

One of the methods to derive streams is to use Strahler order. It has values between 1 and 10, where 1 corresponds to the smallest stream and 10 corresponds to the largest stream. In SAGA, you can use the **Strahler Order** tool, taking **Filled DEM** as input. The next screenshot illustrates how to populate this tool with the new elevation layer with pits removed:

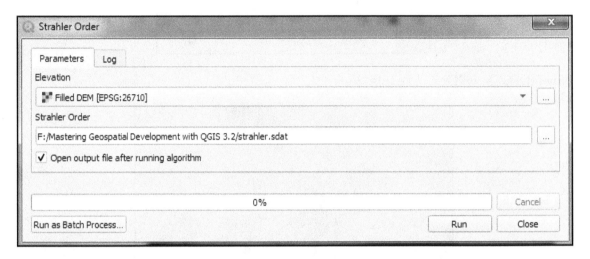

Now, in the **Symbology** tab under **Layer Properties** of the **Strahler Order** layer, use **Singleband pseudocolor** as **Render type**, use **blues** color ramp, and with **Equal Interval** mode, classify into 10 categories, as shown in this screenshot:

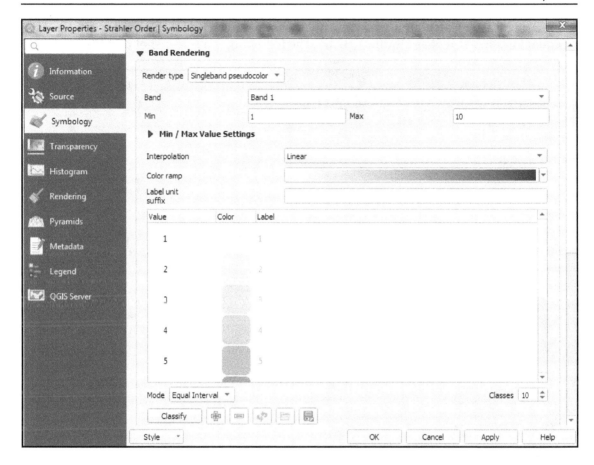

Selecting the streams

Now, using different threshold values of Strahler order, we can decide on which streams are to be considered. For example, if we want to consider only rivers, use a threshold value and see if the resulting stream corresponds to a river in the base map; if not, change this threshold value and check again.

Now, using **Raster Calculator**, we keep only those streams with Strahler order greater than or equal to 7, and save it as **strahler_gt7.tif**. The following is a screenshot of this step:

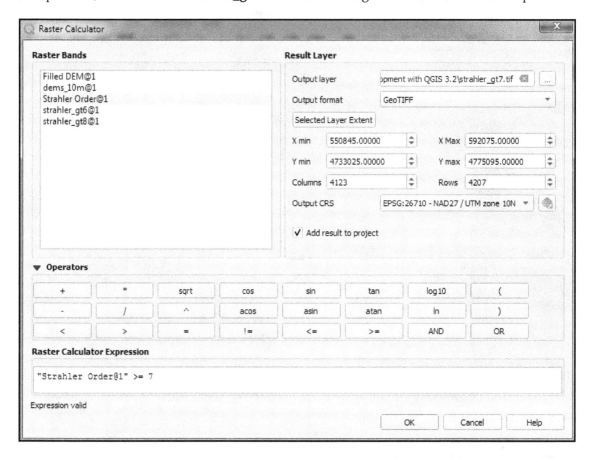

Now, use the **Symbology** tab of the **Layer Properties** of the **strahler_gt7.tif** layer, and keep only values 0 and 1, as shown in this screenshot:

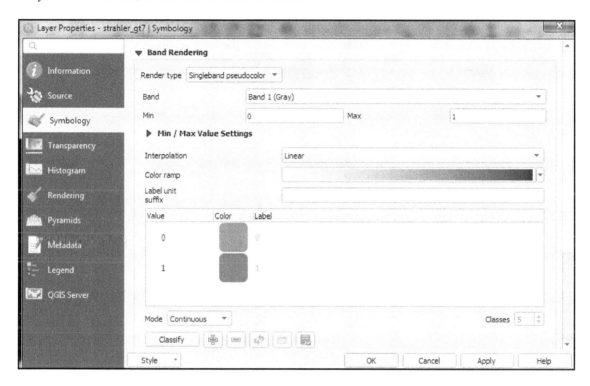

We also need to use a value of 0 for the Additional no data value field under the **Transparency** tab. This step is shown as follows:

Delineating the streams

Now, using the **Channel network and drainage basins** tool of SAGA, we can delineate the streams. Using **Filled DEM** as input, we will get flow direction, channels, and drainage basins. For our case in hand, we will only use channels for further analysis:

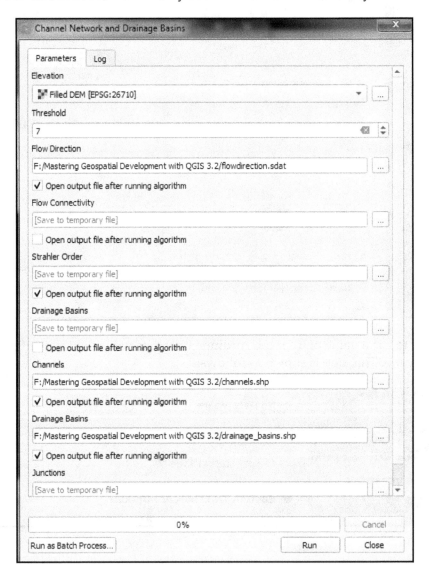

Now, if we activate only the Channels, Gauge (`shapefile`), Filled DEM, and hillshade layers, we can see the streams delineated against a nice backdrop:

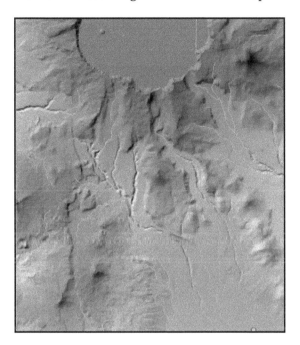

Calculating the upstream area above Fort Klamath

These two grids cover the entire area of the DEM, but we are only interested in evaluating what the watershed looks like along Sun Creek upstream of Fort Klamath. Many rivers are monitored by USGS gauging stations, which can be used as points of interest to delineate the upstream contributing area. However, more often than not, smaller streams typically aren't monitored, even though they are important for local communities. So, we can create arbitrary outlet points that are defined along the stream network. To focus our analyses on Sun Creek, we will make use of the Gauge shapefile and use as our outlet what is often called a pour point. We will use the **Upslope Area** tool to identify the cells that drain through this pour point. In other words, we are going to calculate the watershed above this particular point on Sun Creek. The following screenshot indicates that we need the point coordinate of **Fort Klamath**, which is provided by `gauge.shp`.

Before using any shapefile to calculate the upstream contributing area, it is worthwhile to ensure that every point is located on a grid cell representing the stream network, otherwise the algorithm won't be able to accurately characterize surface flow. To make sure that each point is located on the network, we can use the **r.stream.snap** tool, which will move the point to the nearest stream network:

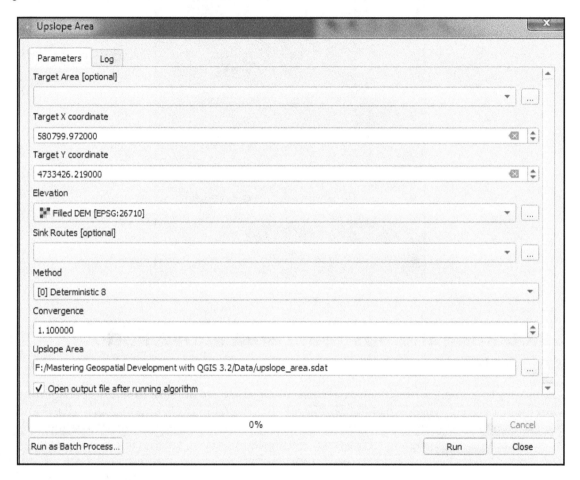

Now, we will have a raster that has a value of 0 for areas outside the catchment, and a value of 1 for areas inside the catchment. We then convert this raster to a vector or shapefile using **Raster | Conversion | Polygonize (Raster to Vector)**. Now, we will remove any 0 value from the attribute table of this new vector (**Vectorized**). The vector layer will now look like this, after we have deselected **Upslope Area**:

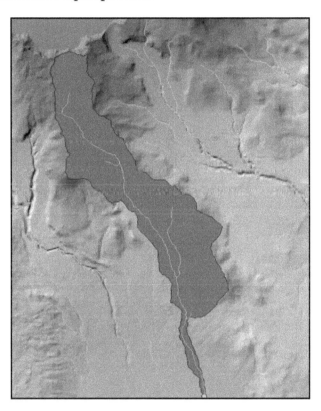

Now, apply a transparent fill and a red stroke color, as shown in this screenshot:

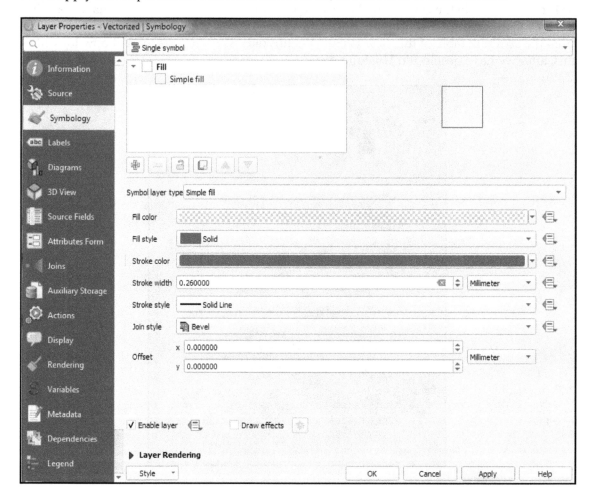

Now, clip the channels layer to the **Vectorized** layer using **Vector | Geoprocessing Tools | Clip**, and populate the interface as shown here:

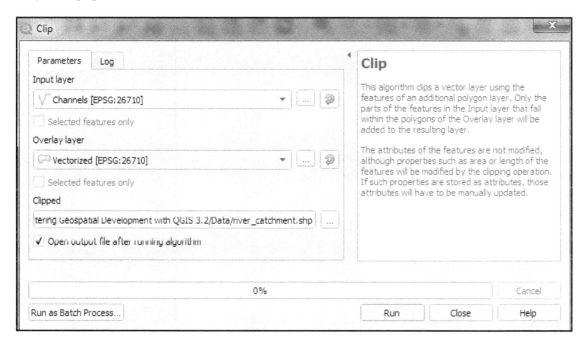

Now, clip the **Filled DEM** raster layer to the **Vectorized** layer using **Raster | Extraction | Clip Raster by Mask Layer**. Now, we will see a new raster layer is added, named **Clipped (mask)**. We now toggle off both the **Filled DEM** and **hillshade** layer, and then copy the style from the **Filled DEM** to the **Clipped (mask)** layer. We do this by right-clicking on the **Filled DEM** layer, and then by selecting **Styles | Copy Styles**, and then again by right-clicking on the **Clipped (mask)** layer, and then by clicking **Styles | Paste Styles**.

Finally, we go to the **Layer Properties** of the raster **Clipped (mask)** and, under the **Transparency** tab, we set 0 as **Additional no data value**. This gives us a delineated streams and watershed raster that will look like this:

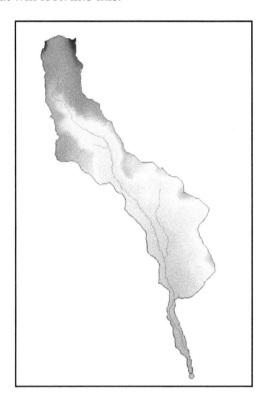

Hopefully, this brief exercise demonstrates the potential applications of SAGA for exploring hydrologic conditions using high-resolution elevation data. Although the final output consists of only models of how water might flow across the surface, this exercise is useful for delineating watersheds from user-specified pour points, and for estimating potential stream networks within this watershed.

Summary

This chapter provided an overview of the structure within the Processing Toolbox, and an introduction to the variety of advanced spatial analysis tools than can be accessed through the Toolbox. You specifically learned how to create a shaded relief map, calculate the least-cost path, evaluate a viewshed, reclassify raster layers, query and combine raster layers, and calculate raster statistics using GRASS algorithms. You then learned how to crop raster layers using a polygon mask, and reclassify, query, and combine raster layers using SAGA algorithms. You learned how to delineate a watershed and extract a vector stream network from a DEM using SAGA algorithms. Perhaps most importantly, we saw how interoperable the native QGIS tools are with the tools executed from within the Processing Toolbox.

Although we explored these tools through hypothetical scenarios to illustrate how these analyses might be applied to real-world questions, it is important to clarify that this chapter is by no means exhaustive in its coverage of the complete suite of tools or their various applications. There are additional powerful algorithms available within the GRASS and SAGA environments. We have also not covered algorithms here within the R and LAStools environments.

In the next chapter, you will learn how to automate geospatial workflows using the graphical modeler within the Processing Toolbox. We will explore the various types of input options and available algorithms, and develop an example model that we can add to the toolbox as a reusable tool. You will also learn how to export models to Python in preparation for the last chapter, which will explore Python scripting within QGIS.

Automating Workflows with the Graphical Modeler

9

This chapter will provide you with an overview of the **Graphical Modeler** (**GM**). First, we will introduce the modeler and explore the various inputs and algorithms available for models. Then, we will demonstrate via step-by-step examples how to develop a model that can be added to the **Processing Toolbox**. We will also cover more advanced topics, including nesting models and executing models interactively. The specific topics that we will cover in this chapter are as follows:

- An introduction to the graphical modeler
- Working with your model
- Executing model algorithms iteratively
- Nesting models
- Using batch processing with models
- Converting a model into a Python script

Introducing the graphical modeler

A typical spatial analysis involves a series of GIS operations, with the output of one operation as the input for the next one, until the final result is generated. Using the graphical modeler, you can combine these individual steps into a single process. The interface to the graphical modeler allows you to visually draw inputs, GIS algorithms, and outputs. The entire analysis is then ready to run as a custom tool within the **Processing Toolbox**. The custom tool will look like other tools in the **Processing Toolbox**. After assigning the inputs and naming the outputs, the entire analysis will run in a single step.

A major benefit of this approach is that the completed analytical workflow can be modified and rerun. This allows stakeholders to understand how changing thresholds or input values affects the results of an analysis. Let's assume that you were assigned the task of developing a site-selection model for a new coffee shop. To match one of the site-selection criteria, you buffered railroads by one kilometer.

However, a stakeholder later asks you how the result would change if the one-kilometer distance was changed to half a kilometer. If you had completed the original analysis with a traditional step-by-step approach, without using a model, you would have to start from scratch to answer this question. However, if you developed this problem as a model, you can simply change the distance parameter in the tool and rerun the entire site-selection model. Similarly, the site-selection model can also be run in a different city or neighborhood simply by pointing to different (but equivalent) input layers. The model can also be shared with others.

Opening the graphical modeler

The graphical modeler can be opened from QGIS Desktop using either of the following two methods:

- By clicking on **Graphical Modeler** under **Processing**
- By enabling the **Processing Toolbox** panel, navigating to **Models | Tools**, and then clicking on **Create new model**

The Processing Modeler opens as a new window. On the left-hand side of the window, there are two tabs: **Inputs** and **Algorithms**. These are used to add both types of element to the modeler canvas, which takes up the remainder of the window. Also to left of the modeler canvas, there are the **[Enter model name here]** and **[Enter group name here]** input boxes to enter the model name and the group name.

The buttons for managing models can be found above the **Inputs** and **Algorithms** tabs, as shown in the following screenshot:

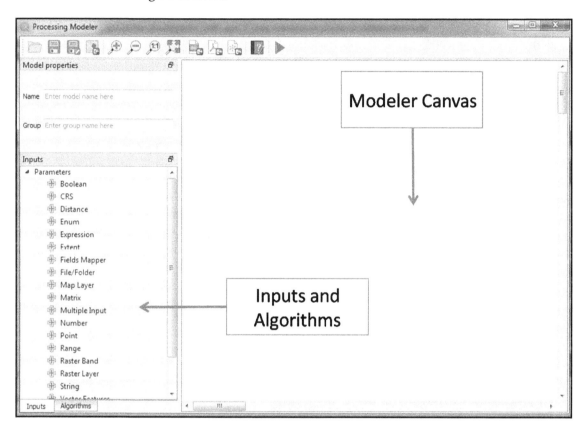

The window itself is called **Processing Modeler**, and not Graphical Modeler.

Configuring the modeler and naming a model

Before starting a model, it is good practice to configure the modeler. Models are saved as XML files with a .model3 extension. When you save a model, QGIS will prompt you to save the model file to the models folder. You can set the location of the models folder by clicking on the 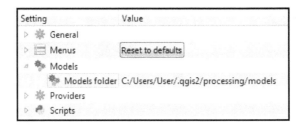 icon under **Processing Toolbox** in QGIS Desktop. Under the **Models** properties section of the **Processing options** window, you can specify the location of the **Models folder**. Double-click on the default folder path, and the browse (ellipsis) button will appear, allowing you to add a different location:

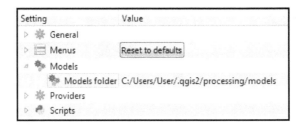

To demonstrate the basics of using the graphical modeler, we will use a simple example that identifies riparian tree stands in Alaska. It will have three inputs and two algorithms. First, we will give our model a name and a group name. For this example, as shown in the following screenshot, we have opened the graphical modeler and named the model Riparian Trees and the model group Landcover. This is the group and the name by which the model will be displayed within the processing toolbox:

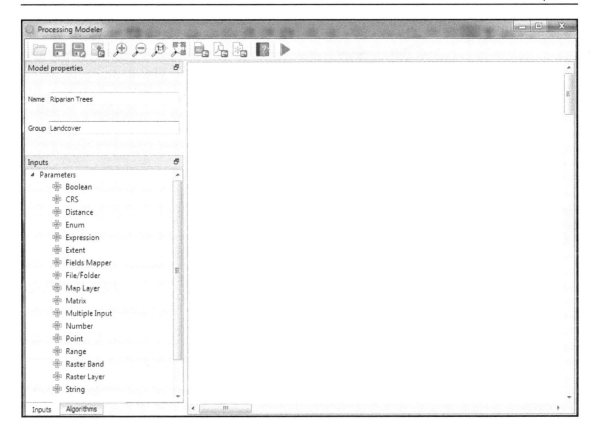

In the Processing Modeler window, we write `Riparian Trees` in the textbox corresponding to **Name** and `Landcover` for **Group**.

Then, we will click on the save button ( ). The **Save Model** dialog will open, defaulting to the `models` folder. Here, you need to choose a name for the `*.model3` file. We are naming it `RiparianTreeClipper.model3`.

> The model name and the group name must be set before the model can be saved.

The following screenshot shows the **Save Model** dialog box:

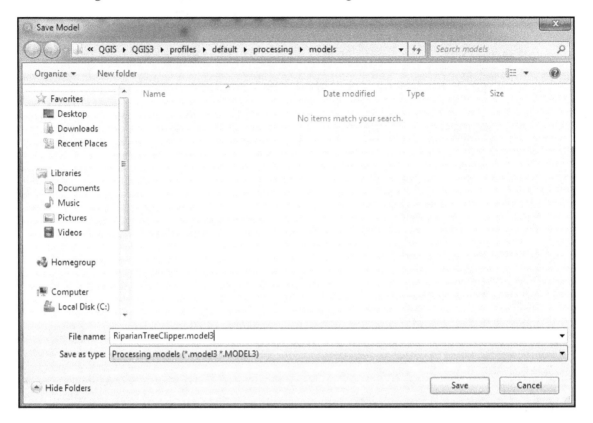

If the models are saved to the models folder, they will appear as model tools in the **Processing Toolbox** panel. Once a model has been named and saved to the models folder, it will appear under its group in the Processing Toolbox. Again, the model will appear with the name that was entered into the graphical modeler rather than the name of the * .model3 file. Models can be saved outside the models folder, but they won't appear in the **Processing Toolbox** panel.

The next screenshot shows the **Processing Toolbox** with the `Models` section expanded with the `Riparian Trees` model:

In the **Processing Toolbox,** under the **Models** category with the **Landcover** group, we can find **Riparian Trees** model.

Working with your model

In this section, we will have a brief overview of how to start working on your models. We will now add inputs and algorithms, and see how you can run your models.

Adding data inputs to your model

In this first example, you will identify the portion of a forest within 100 meters from streams (riparian trees). To begin a model, you will need to define the inputs. The graphical modeler will accept the following:

- Boolean
- Extent
- File
- Number
- Raster layer
- String
- Table
- Table field
- Vector layer

To add an input, either double-click on the appropriate category from the **Inputs** tab or drag the input on to the modeler canvas. The **Parameter Definition** dialog will open. Give the parameter a name and fill out any other details, which change depending on the input that is chosen. When an input parameter is defined and added to the model, it is essentially a conceptual parameter. It will not actually be connected to a GIS data layer until you are ready to run the model.

For this example, we will add a vector layer. Double-click on **Vector Layer** under the **Inputs** tab. We will specify the geometry of the vector data as **Line** and classify it as **Mandatory**:

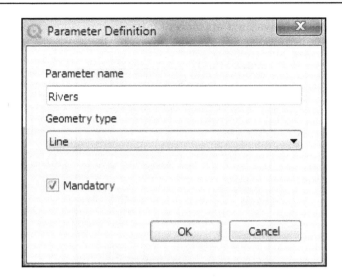

Once you click on **OK**, the input object is added to the modeler canvas. All the objects in the modeler canvas can be selected with a mouse click and dragged to reposition them. Clicking on the pencil icon of an input will open the **Parameter Definition** dialog so that changes can be made to it. Clicking on the close button (**X**) will delete the input from the model.

For our example, we will add a second vector layer. Trees (forest) is added as a required polygon layer. Finally, we will add a number input. This will allow us to expose the buffer distance value as an input that can be changed when the model is executed. It will be named Buffer distance and it will be given a default value of 100, since 100 meters is the distance that we initially want to use.

In the graphical modeler, distances are expressed in coordinate reference system units.

The following screenshot shows the **Parameter Definition** dialog box, as described previously:

The following screenshot shows the model with the two vector layer inputs and a number input:

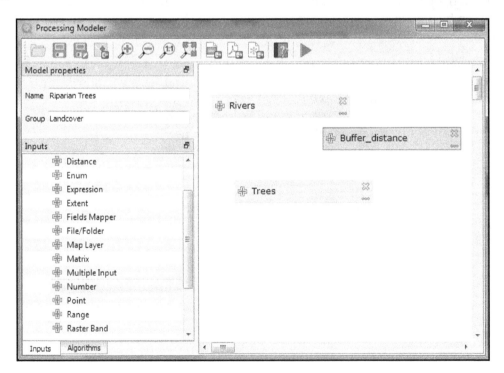

As you can see, this is a model with three inputs.

Adding algorithms to your model

Algorithms are added to the graphical modeler in the same way as inputs. Find the algorithm from the **Algorithms** tab, and either double-click on it or drag it on to the modeler canvas. You can search for tools as you would in the **Processing Toolbox**. Type the name into the search box at the top of the **Algorithms** tab:

You can see the **Algorithms** tab with `buffer` being used as the search term.

The algorithm dialog will look very similar to how it would if you were running it from the **Processing Toolbox**. There are inputs, tool parameters, and outputs. However, there are some important differences because the graphical modeler is a self-contained universe of data inputs. The differences are as follows:

- Input layers are limited to those that have been added to the model.
- Output can be left blank if it is an intermediate result that will be used as an input for another algorithm. If the output is a layer that needs to be saved, enter the name of this layer into the textbox. When naming an output layer, you won't actually need to provide an output filename. This will be done when the tool is run. Instead, you just need to enter the name of the layer (for example, `stream buffer`).
- Numerical values or string-value parameters can be entered as numbers or strings. They can also be chosen from other inputs of the number or string type.
- The fields of an attribute table (or another standalone table) can be specified by typing the field name or by using the **Table field** input. These fields will be chosen when the model is run.
- **Parent algorithms** is an additional parameter found only in tools that are run from the graphical modeler. It allows you to define the execution order of the algorithms. Setting an algorithm as a parent forces the graphical modeler to execute this parent algorithm before the current algorithm can be run. When you set the output of one algorithm as the input for the next one, it automatically sets the first algorithm as the parent. However, in complex models, there may be several branches, and it may be necessary for an operation in a separate branch of the model to be completed before another operation can run.
- Again, in this example, we will be buffering streams by 100 meters and then clipping trees by that buffer layer. The first algorithm that we will add is **Fixed distance buffer** under **Saga | Vector General**. Double-click on the tool from the **Algorithms** tab, and the tool dialog will open. The tool will be filled as shown in the following screenshot. Notice that instead of setting an explicit buffer distance, the buffer distance input is being used; we get this by clicking on the symbol 🧩 and then selecting **Model Input**. Also, note that no output is named, since this output will be considered an intermediate dataset:

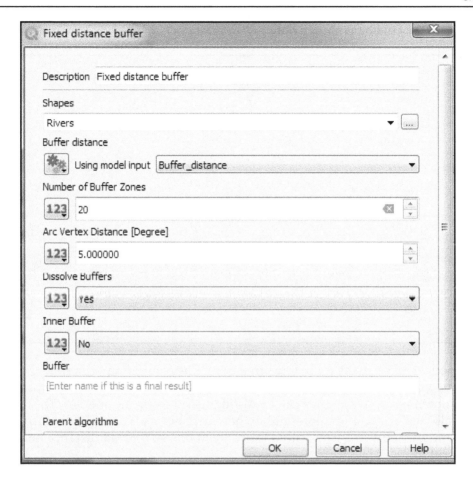

Next, we'll add the **Clip** tool to the model, using the following parameters:

- Set the **Input layer** field to **Trees**
- Set the **Overlay layer** field to **'Buffer' from algorithm 'Fixed distance buffer'**
- Type `Riparian Trees` under **Clipped**
- Finally, click on **OK**

The completed tool dialog is shown here:

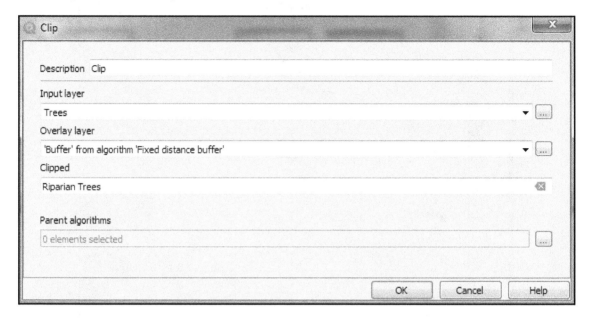

The final model looks like the following screenshot. The connecting lines show how elements are connected in the workflow. The input, output, and algorithm elements have different-colored boxes so that they can be distinguished from one another. The algorithm boxes will also include an icon representing the source library. For example, **Clip** is a QGIS algorithm and has the arrow icon within the element box, and **Fixed distance buffer** is a SAGA algorithm:

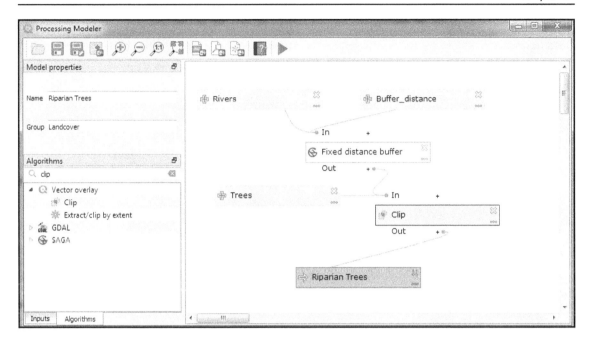

Running a model

The model can be run either from the **Processing Modeler** window or from the **Processing Toolbox** panel. To run a model from the **Processing Modeler** window, click on the **Run model** button (▷). To run a model from the **Processing Toolbox** panel, first save and close the model. Then, find the model by navigating to **Processing Toolbox** | **Models**, right-click on it, and choose **Execute...** from the context menu. In our example, the model will be found in the **Landcover** group:

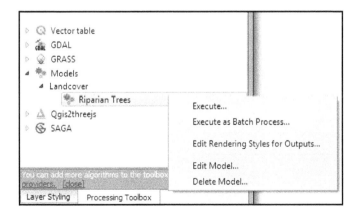

The model dialog will open with the listed inputs. For the data layer input, you can choose data loaded into QGIS by using the drop-down arrow or you can use the browse button (

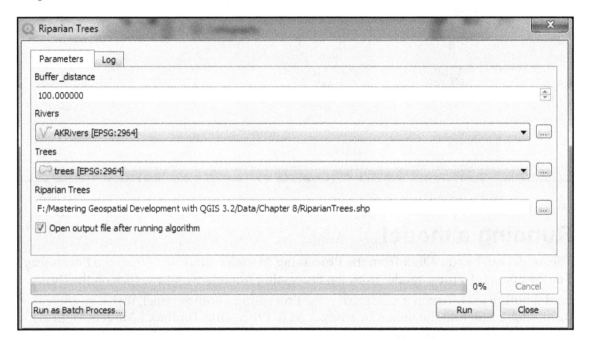

) to locate the data on disk. For this example, we are using the AKrivers.shp and the trees.shp sample data. The **Buffer_distance** field is set to 100, since this was the default value set for the number input. For the output, you can choose to have the layer as a temporary one, or choose a location and filename for it. Here, the data is being saved as a shapefile. Click on **Run** to execute the model:

As the model runs, the dialog will switch to the **Log** tab, which provides output as it runs:

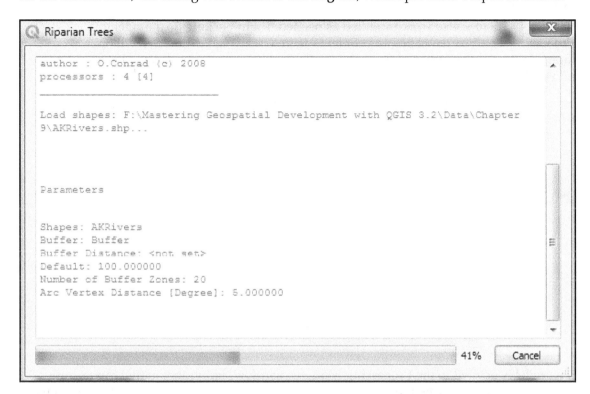

All the model files and data inputs discussed in this chapter are included with the *Mastering QGIS* sample data.

Editing a model

Existing QGIS models can be modified as needed. Right-clicking on a model in the **Processing Toolbox** panel opens a context menu. Choosing **Edit model** will open the model in the **Processing Modeler** window. The model can also be deleted here by clicking on **Delete model...**:

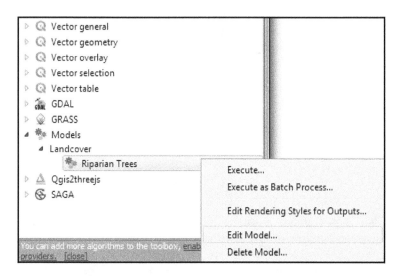

If a model is opened in the **Processing Modeler** window, individual model inputs and algorithms can be modified. As we mentioned in the *Adding inputs* section, clicking on the pencil icon of a model's input will open the **Parameter Definition** dialog so that changes can be made. Clicking on the close button (**X**) will delete the input from the model:

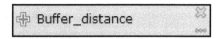

From the modeler canvas, information about algorithm parameters can be exposed by clicking on the + signs above and below an algorithm. This is a convenient way to see algorithm parameters without opening each algorithm. Right-clicking on an algorithm opens a context menu, as you can see in the following screenshot.

Clicking on **Remove** deletes the algorithm from the model, as long as there are no other algorithms depending on its output:

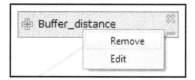

If you attempt to delete an algorithm in the middle of a workflow, you will see the following message. The dependent downstream elements will have to be deleted prior to deleting the algorithm:

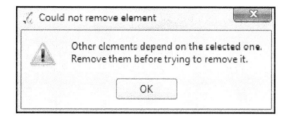

Clicking on **Edit** from the algorithm context menu opens the algorithm dialog so that changes can be made to the model. After editing an algorithm, the connections to other model elements in the canvas will be updated. The algorithm parameters exposed by clicking on the + signs above and below the algorithm will also be updated:

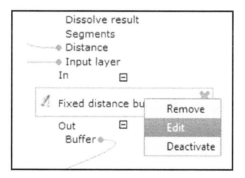

Clicking on **Deactivate** from the algorithm context menu will deactivate the algorithm and all algorithms downstream that depend on that algorithm. An algorithm can be reactivated at any point by right-clicking on it and choosing **Activate**. When you do this, any other downstream algorithms that were deactivated earlier will have to be individually reactivated.

Documenting a model

Model help can be written for any model by clicking on the Edit model help button (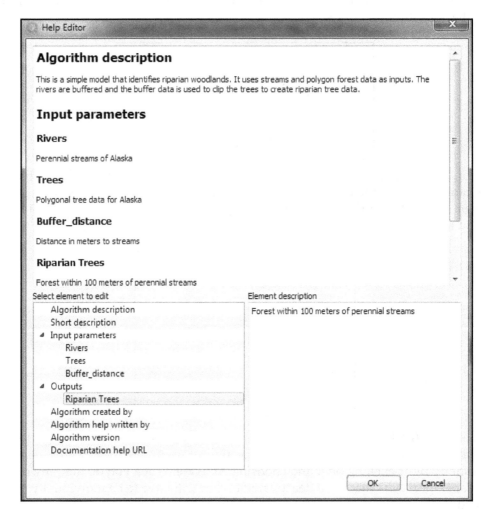) within the **Processing Modeler** window. This will open the **Help editor** window, which has three panels. At the top is an HTML page with placeholders for the **Algorithm description**, **Input parameters**, and **Outputs** sections. At the bottom-left corner, there is an element selection box, and there is a box for entering text at the bottom-right corner. To edit an element, select it in the **Select element to edit** box. Once it is selected, use the **Element description** box to type a description or necessary documentation. Click on **OK** when finished:

This help information will then be available on the **Help** tab when the tool is in execution mode:

Saving, loading, and exporting models

Models can be saved at any time by clicking on the **Save** button (![save icon]) in the **Processing Modeler** window. It is best to save early and often when working on a model. As we mentioned in the *Configuring the modeler and naming a model* section, the first time a model is saved, you will be prompted to name the model file. Subsequent saves update the existing `*.model` file. There is also a **Save as** button (![save as icon]), which can be used to save a new version of a model.

Models that are not saved to the **Modeler folder** can be opened using either of the following two ways:

- By opening the **Processing Toolbox** panel, navigating to **Models** | **Tools**, and then double-clicking on **Add model from file.**
- By using the **Processing modeler** window and clicking on the **Open model** button (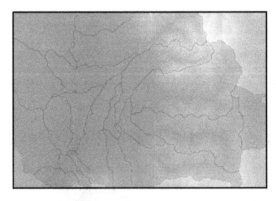).

- In either case, navigate to the `*.model` file.

Models can also be exported as image files. This is useful if the workflow needs to be presented or included in a report. To export a model, click on the **Export as image** button (

) in the **Processing Modeler** window. The model will be saved as a PNG file.

Now that you know the basics of modeling and you have already added an algorithm to it, we will now move on to the execution phase.

Executing model algorithms iteratively

Models, like all QGIS algorithms, can be executed iteratively. Here, we will create a new QGIS model: `DEMs_Clipped_to_Watersheds.model3`. We will use two inputs, a DEM covering Taos, New Mexico and a watersheds polygon layer for the area.

The `elevation.tif` and `watersheds.shp` sample data will be used:

Input data: watersheds and a DEM

Now create this model with a vector layer input named **Watersheds** and a raster layer input named **DEM**. Then, use **Clip raster with polygon** by typing it in the **Processing Toolbox** search area. After we open this by double clicking, use **DEM** for **Input** field and **watersheds** for **Polygons** field and name the **Clipped** region as Watershed DEM:

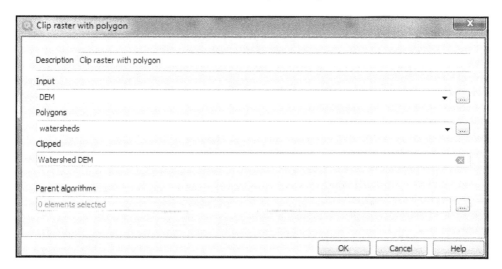

The model will now look as follows:

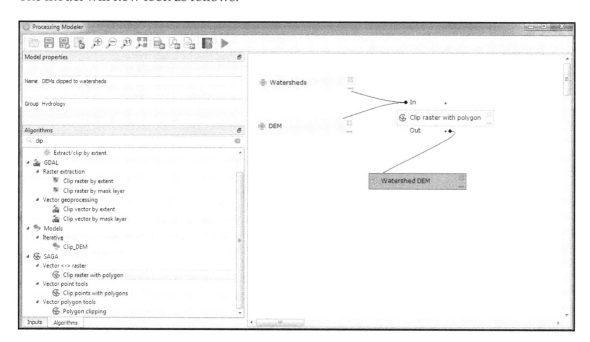

This model uses the **Clip raster with polygon** tool to clip the DEM to the watersheds. There are 21 watersheds covering this area. If the model is run normally, it will clip the DEM to the extent of all the watersheds and produce one output elevation raster. However, if the

Iterate over this layer button (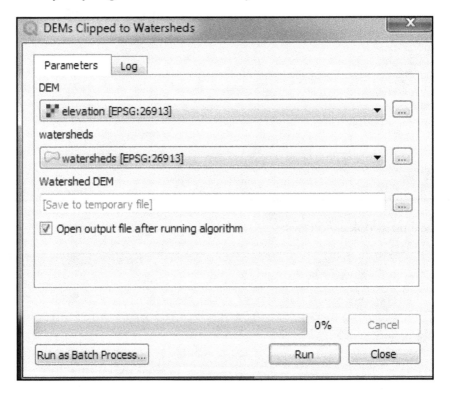) is clicked (see the following screenshot), the model will cycle through each feature in the watershed layer and output a DEM that covers each individual watershed. This will result in 21 individual elevation rasters. This sort of automation is very easy to generate and can save you a lot of time:

The following screenshot shows the resulting 21 DEMs, which are clipped to the individual watersheds:

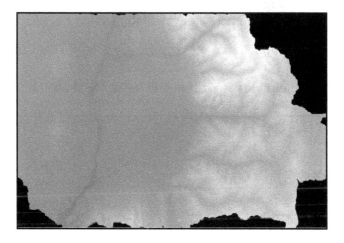

Nesting models

As we previously covered, when a model is saved to the `models` folder, it will appear in the `Models` category of the **Processing Toolbox** panel. What we didn't mention earlier was that it will also appear in the **Algorithms** tab of the **Processing Modeler** window. This means that a previously written model can be used as an algorithm in another model.

Models won't appear as algorithms if some of their component algorithms are not available. This can happen if an algorithm provider is deactivated in **Providers**, and you can find this by navigating to **Processing | Options**. For example, if you have used a SAGA tool in a model but have subsequently deactivated SAGA tools, that model will not be available. As long as all the algorithms in a model are visible in the **Processing Toolbox** panel, a model will be available as an algorithm.

To demonstrate this feature, we will build on the model that we used in the previous section. The model clipped elevation data by watershed boundaries. With a DEM, you can generate a metric called **Topographic Wetness Index** (**TWI**). Now, let's create a QGIS sample model (`TWI_from_DEM.model3`) for calculating TWI.

1. First, we create a model, `TWI_from_DEM.dem3`, by first adding a raster layer and naming it **DEM**
2. Add the **Catchment area** algorithm and select **DEM** for the **Elevation** field:

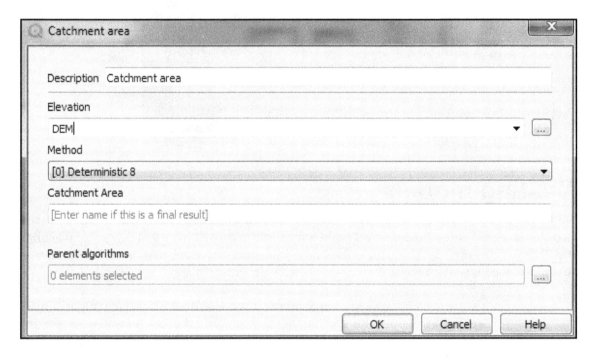

3. Then, add the **Slope**, **aspect**, **curvature** algorithm under **SAGA** and use **DEM** as the input for **Elevation**

4. Now add `Topographic wetness index (twi)` and use slope raster that you got using algorithm **Slope, aspect, curvature,** and use catchment area raster file that you got using the algorithm **Catchment area.** Please have a look at the screenshot below for clarification:

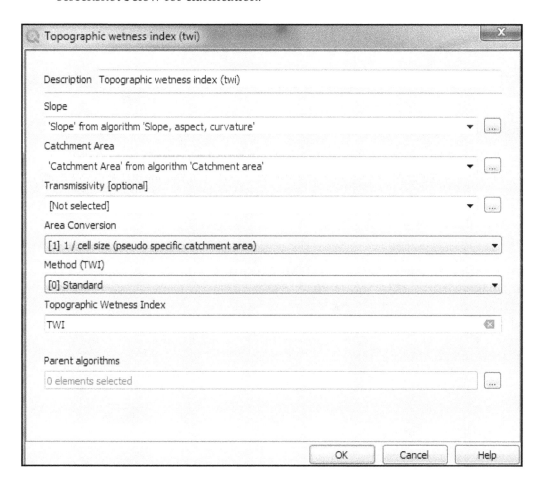

The `TWI from DEM` algorithm then looks like this:

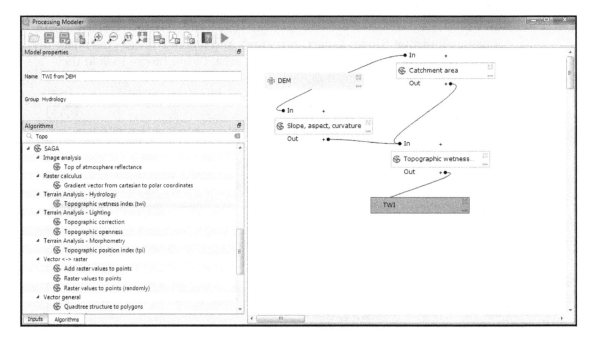

First, we will create a new copy of the **DEMs Clipped to Watersheds** model using the Save as button. We will name this new model file `TWI_for_watersheds.model3`. The **TWI from DEM** model is located in the **Algorithms** tab and is added as an algorithm to our new model. (Remember that models need to be saved to the `models` folder to appear as algorithms.) You will notice that the model icon in the modeler canvas for the **TWI from DEM** algorithm identifies the algorithm as a model:

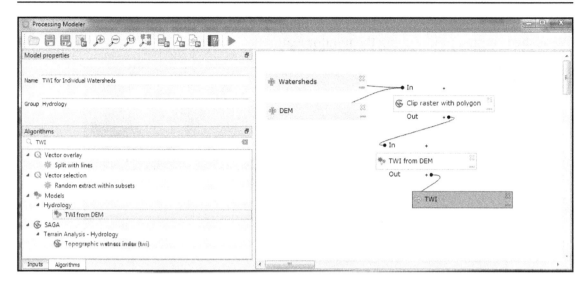

The DEM will be clipped to the Watersheds layer, and the clipped DEM will be the input to the TWI from the DEM algorithm. This will create one output, a TWI raster covering the watersheds. However, if the **Iterate over this layer** setting is used, the DEM will be clipped to each of the 21 watersheds and the TWI will be calculated for each. This will use both a nested model and the iterate feature in the same model:

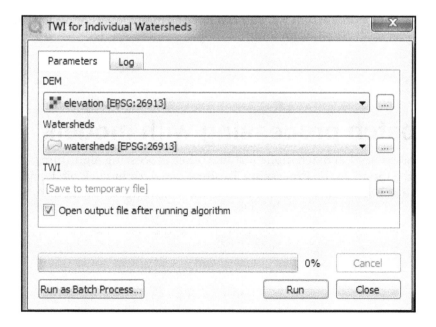

The following screenshot shows the output of the TWI for the **Individual Watersheds** nested model using the iterator feature:

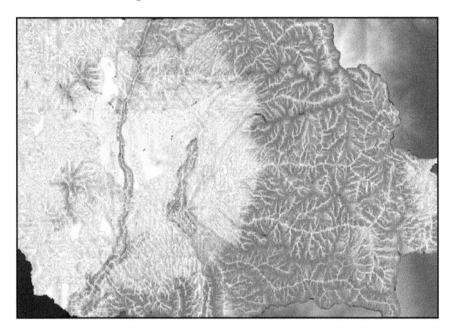

 As of QGIS Version 3.4, you can have as many levels of nested models as you wish. There is no limit!

Using batch processing with models

Models can also be used in batch mode just like other processing algorithms. To do this, simply locate the model in the **Processing Toolbox** panel, right-click on it, and choose **Execute as Batch Process...**, as shown in the following screenshot:

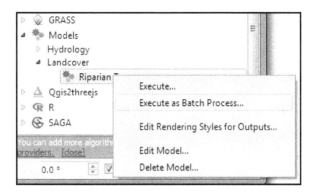

The tool will operate like any other geo algorithm in QGIS. You can click on **Add rows**, **Delete rows**, and **Run** when you are ready. With this method, the model can be utilized on datasets from different geographies. This technique is also useful in cases where you have to repeat several geoprocessing steps on a collection of files. You can do this by adding different steps by clicking on ⊕ and specifying **Buffer distance** and files for **Rivers** and **Trees**:

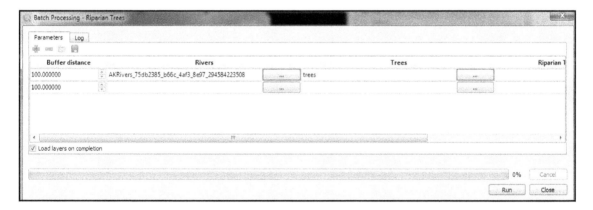

Converting a model into a Python script

In **QGIS 3.4**, the option to directly save a model as Python script is not available. `*.model3` files are now stored as XML, which allows you to nest models within models, and allows algorithms and inputs to be dragged and dropped on to the modeler canvas. However, since models are algorithms, they can be executed from the Python console. This topic is covered in `Chapter 11`, *PyQGIS Scripting*.

Summary

In this chapter, we covered automating workflows with the QGIS graphical modeler. We showed you how to set up, edit, document, and run a model. You learned how to add inputs and algorithms to models. We also covered how to execute models iteratively, nest models within models, and run them in batch mode. With what has been covered up to this point, you should understand how to work with a variety of vector, raster, and tabular data. You should also be well-versed in the geoprocessing and analytical capabilities of QGIS.

In the next chapter, we will switch from conducting analyses with the graphical modeler and the Processing Toolbox to expanding the functionality of QGIS with Python. In `Chapter 10`, *Creating QGIS Plugins with PyQGIS and Problem Solving*, you will learn how to create a QGIS plugin from scratch. The chapter will begin with a primer on PyQGIS. You'll learn where you can get API information and other PyQGIS help. We will then explore the plugin file structure and the available functions. The chapter will conclude with a simple step-by-step example of writing a QGIS plugin. This will also include information on debugging your code.

10
Creating QGIS Plugins with PyQGIS and Problem Solving

In this chapter, we will work with Python on QGIS. We will look at the sources that provide the information on APIs, and also how you can connect with the core QGIS community. This chapter focuses on the basic information necessary to start developing a QGIS plugin.

This chapter will cover the following topics:

- Webography- Where to get help to solve your PyQGIS problems
- The Python Console
- My second PyQGIS code snippet-looping the layer features
- Exploring iface and QGis
- Exploring a QGIS API in the Python Console
- Creating a plugin structure with Plugin Builder
- A simple plugin example
- Setting up a debugging environment
- Debugging session example

Webography - where to get API information and PyQGIS help

One of the characteristics of most free software projects is that their documentation is freely available and can be used for learning. QGIS is one of the best-documented projects, thanks to training material, a coding cookbook, and the automatic documentation of its **Application Programming Interfaces (APIs)**.

In this chapter, we will focus on main resources that are available on the web to learn how to script QGIS and how to solve your scripting problems.

PyQGIS cookbook

The main resource is a community- and content-driven cookbook that gives a general introduction to scripting QGIS. You can find this documentation at http://www.qgis.org/en/docs/index.html.

You have to choose the QGIS target version of your plugin and then choose the **PyQGIS Developer Cookbook** link. If you are interested in the latest QGIS APIs, you have to choose the testing version, or go directly to https://docs.qgis.org/3.4/pdf/en/QGIS-3.4-PyQGISDeveloperCookbook-en.pdf

If you need a copy on your system, you can also download a PDF version of the cookbook from https://docs.qgis.org/testing/pdf/en/QGIS-testing-PyQGISDeveloperCookbook-en.pdf.

For example, if you need the documentation for older version such as 2.18, you have to direct your browser to https://docs.qgis.org/2.18/pdf/en/QGIS-2.18-PyQGISDeveloperCookbook-en.pdf.

However, if you need documentation for a version under development, you can visit http://docs.qgis.org/testing/en/docs/pyqgis_developer_cookbook/.

Once you open the PDF documentation page, you can choose your preferred translation.

 QGIS even versions (only the decimal number, for example, 3.6, 3.4, 2.6 and so on) are always stable versions. Odd versions are always developing versions (for example, 2.3 or 2.5). Odd versions are generally known as testing versions.

API documentation

APIs are the doors that the components of a software program use. As a program evolves, the APIs can change affecting, in our case, all the plugins that directly use them.

There were many changes as QGIS jumped from 2.*x* to 3.*x* as it was the case with the jump from version 1.8 to 2.0, but moving from one version to another, there are new APIs added and others are deprecated because of the normal development life cycle of complex software. Generally, deprecation and new APIs are added to enable new features (for example, custom validity check for layout added in version 3.6) or due to code refactoring (refer to `http://en.wikipedia.org/wiki/Code_refactoring`).

So, depending on API changes and on what QGIS version you want to integrate your code into, you should use one API set or another, or better, write code that can be executed in multiple QGIS versions.

API documentation is automatically generated from the QGIS code and can be found at `http://qgis.org/api/`, where you can look for the class that you need. Here, you can find API documentation for all the versions of QGIS.

For example, if you need to know public methods of the QgsVectorLayer class in, visit latest development version, `http://qgis.org/api/classQgsVectorLayer.html`. But if you want to know the public methods of this class in one of the long term release versions, QGIS 3.4, you can refer to `https://qgis.org/api/3.4/classQgsVectorLayer.html`.

It should be mentioned that QGIS is mainly written in C++ and its API documentation follows C++ notation. Most of the methods of the QGIS classes are available as a Python bind. The way to discover whether a method is exported to Python is to test it in the QGIS Python Console or read about its class documentation in the QGIS Python Console, as described in the next chapter.

The QGIS community, mailing lists, and IRC channel

One of the advantages of the open source project is that you can talk directly to other developers and frequently with the core developers of the project. QGIS has three official ways to support development and problem resolution.

Mailing lists

All official mailing lists are listed at `http://qgis.org/en/site/getinvolved/mailinglists.html`.

There are two extremely important lists from the user's point of view: one for developers (QGIS core and plugin developers) and the other for users. Depending on your profile, choose one list:

- **Developer list**: For this, refer to `http://lists.osg-+eo.org/mailman/listinfo/qgis-developer`
- **User list**: For this, refer to `http://lists.osgeo.org/mailman/listinfo/qgis-user`

These lists can be read and searched for using an online service at `http://osgeo-org.1560.x6.nabble.com/Quantum-GIS-f4099105.html`, where you can also find lists of other QGIS sub-projects or the mailing lists of several other local QGIS user groups.

If you don't find your nearest local QGIS user group, first ask in the user or developer mailing lists, and, depending on the answer, try to create a new QGIS local group and announce it to the community.

IRC channel

Internet Relay Chat (**IRC**) is a fantastic way to get real-time support from users and developers. Remember that this help is always voluntary and the answers depend on your politeness and the available time of the connected users.

You can connect to the #qgis channel at the `http://www.freenode.net` server with your preferred IRC client via `http://webchat.freenode.net/?channels=#qgis`. Alternatively, you can connect using the gitter service via `https://gitter.im/qgis/QGIS`.

The philosophy of IRC problem solving is condensed in the first chat message sent to you by the #qgis channel; it will be *Don't ask to ask, just ask and hang around a while to see if someone answers. Please refer to* `http://osgeo.pastebin.com/` *instead of pasting more than five lines.*

The Stack Exchange community

Technical social networks such as Stack Exchange have a GIS sub-project that can be accessed from `http://gis.stackexchange.com`. Here you can look for problems reported by other users about QGIS and the answers that are given by other users or directly by QGIS core developers.

Messages relating to QGIS can be found by looking for the qgis tag, for example, `http://gis.stackexchange.com/?tags=qgis`.

Sharing your knowledge and reporting issues

In Stack Exchange, the IRC channel, and the mailing lists, you can actively support other users who have problems that you have have already solved.

An important way to support a QGIS project, other than funding it, is by reporting bugs that are packaged with a detailed use case and data that allows others to replicate the problem. This will speed up bug fixing.

There are two kinds of issues: those that are related to QGIS or its core plugins (such as **Processing**) and those that are related to third-party plugins that you can install with **Plugin Manager**.

To report a QGIS issue or a core plugin issue, you need an OSGeo account, created at `https://www.osgeo.org/cgi-bin/ldap_create_user.py`, using which you can log in to the QGIS redmine bug tracker to report issues at `http://hub.qgis.org/projects/quantum-gis/issues`.

> Beware! Look in the issues list first to make sure your issue hasn't already been reported.

A good guide to reporting a QGIS issue is available at `https://www.qgis.org/en/site/getinvolved/development/bugreporting.html`

Reporting a third-party plugin issue depends on the plugin developer and where he/she decides to host the bug tracker. You can find this information in the Plugin Manager, as shown in the following screenshot:

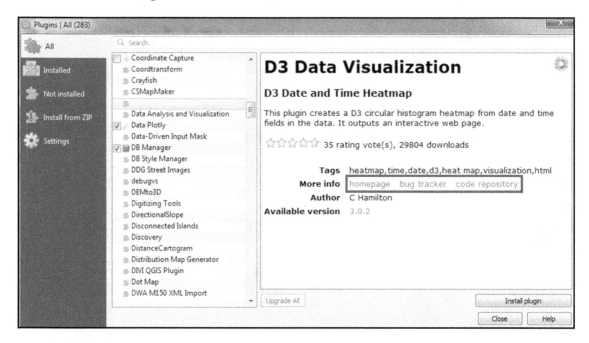

If you are looking for the link tracker, it will be found in the area marked by the red box in the second link, **tracker**. Other useful links can be found in the area marked in red. The first link points to the **homepage** plugin, where the plugin is described. The third link points to the the **code repository** of the plugin. In the QGIS plugin central repository, every plugin is accepted only if it has a lease tracker and a code repository.

The Python Console

The Python Console is a wonderful instrument to explore and learn PyQGIS. It's available in every QGIS installation and can be opened by selecting **Python Console** in the **Plugins** menu.

The Python Console is a dockable interface, and like all dockable interfaces, you can change its position inside QGIS or separate it. You can try moving the console by dragging and dropping it.

The console is shown in this screenshot:

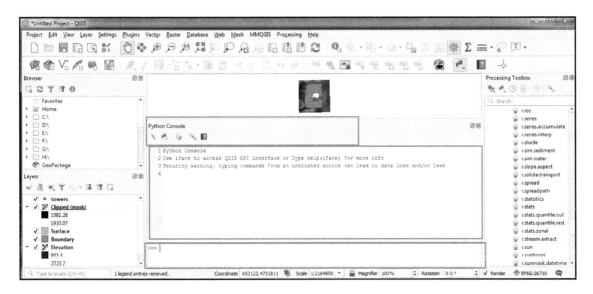

As you can see, the console is composed of a button toolbar marked by the red box. The bigger **Python Console** is marked by the upper-right red box and is where all the command results are shown, and finally a bottom command line, marked by the bottom-right red box, is where you can edit commands. The default position of these graphical components can vary with the QGIS version.

Here, we describe how to test code interactively and explore the PyQGIS classes. However, we will not explain all the possibilities of the Python Console. These are well documented and you can find them at `http://docs.qgis.org/testing/en/docs/pyqgis_developer_cookbook/intro.html#python-console`.

Getting sample data

To continue experimenting with PyQGIS, we need a test dataset. The QGIS project has training material and sample data that we'll use in our snippet.

The QGIS sample data can be downloaded from `https://qgis.org/downloads/data/qgis_sample_data.zip`.

My first PyQGIS code snippet

To break the ice, we will create our first PyQGIS code to show the unique ID of a selected layer loaded in QGIS:

1. To start with, we will first load the layer, `airports.shp`, which is available in the `shapefiles` directory of `qgis_sample_data`

2. After it has been loaded successfully, select it in the list of layers; doing this will make it the active layer for QGIS

3. Then we get the reference of the layer object by writing the following line in the Python Console command line:

   ```
   layer = iface.activeLayer()
   ```

4. After editing the code and hitting Return, the edited command code is shown in the console.

5. The reference of the current active layer is archived in the variable named layer. Now, we can show the layer ID by typing the following code command in the command line after pressing Return:

   ```
   print(layer.id())
   ```

The output will display something similar to:

```
airports_5f63019a_84a3_4b58_9f84_d3282bfc0a21
```

My second PyQGIS code snippet - looping the layer features

In this section, we'll introduce how to loop in Python and how to apply loops to explore the content of the layer loaded in the previous paragraph. Write the following snippet in the Python Console, taking special care with the code indentation:

```
for feature in layer.getFeatures():
    print("Feature %d has attributes and geometry:" % feature.id())
    print(feature.attributes())
```

This will print a pattern like the following:

```
Feature with id 68 has attributes and geometry:
[69, 9, 12.0, 'HAINES', 'Other']
```

The `layer.getFeatures()` method returns an object that can be iterated inside a `for` Python instruction, getting a QgsFeature instance for every loop. The `feature.attributes()` method returns a list (inside the brackets, `[]`) of the integer and Unicode strings (the u' values). The `feature.geometry()` method returns `QgsGeometry`, which is converted in `QgsPoint` to be printed as a tuple (inside the `()` parenthesis) of coordinates.

It is strongly recommended that you explore the preceding classes. You can also practice by referring to the documentation at `http://qgis.org/api/`. Start by exploring the `Qgis bInterface` and `QGis` classes.

Indentation is an important part of the Python language; in fact, nesting the code in Python is done using indentation, as specified in the standard followed globally. You can find this at `http://legacy.python.org/dev/peps/pep-0008/`.

Exploring iface and QGis

The `iface` class used in the preceding snippets is important in every PyQGIS code; it is used to access most graphical QGIS components, from displayed layers to the toolbar buttons.

The `iface` class is a Python wrapper for the C++ class, **QgisInterface**, which is documented at `http://qgis.org/api/classQgisInterface.html`. Most QGIS classes have a **Qgs** prefix. Some special classes can have the Qgis or QGis prefixes. The Qgs is the Qt namespace registered by Gary Sherman, the QGIS creator, so **Qt (Q)** and **Gary Sherman (gs)**.

The most common use of the iface class is to get a reference of the canvas where maps are displayed:

```
canvas = iface.mapCanvas()
```

The class can also be used as a shortcut to load raster or vector layers; for example, loading the raster, `path/to/my/raster.tif`, and naming it `myraster` in the legend panel. This can be done by typing the following command:

```
iface.addRasterLayer("path/to/my/raster.tif", "myraster")
```

Pay attention when writing paths with Windows. A path string, such as `C:\path\to\raster.tif`, has the special escape character, `\`, so rewrite it by double escaping `C:\\path\\to\\raster.tif` or using the Unix notation, `C:/path/to/myraster.tif`, or notify Python with a raw string adding an r as in `r"C:\path\to\raster.tif"`. Generally, it's good practice to create path strings using a Python library such as `os.path`.

`QGis` is another class that contains some useful constants, such as a QGIS version or some default values.

We can find out the QGIS version name running on our system by typing in the following command:

```
from qgis.core import Qgis
print(Qgis.QGIS_RELEASE_NAME, Qgis.QGIS_VERSION_INT)
```

For example, if the output is `Madeira 30404`, then this value represents the version name and the version integer representation (which is version 3.4). This is useful for programmatically creating a plugin that can run on different QGIS versions. The following snippet helps to distinguish the code among them:

```
if Qgis.QGIS_VERSION_INT < 30400:
<here the code compatible with older version>
else:
<here the code compatible with version higher or equal to 3.4>
```

Exploring a QGIS API in the Python Console

The QGIS APIs can be browsed in the documentation web page, but if you want to access the documentation directly in the Python Console, you can use some useful Python commands. The `help` command shows a synthesis of the API information available in the web documentation. Try to edit the Python Console with this command:

```
help(iface)
```

The console will show all the methods of the `QgisInterface` class and a synthetic example of how to use this in Python syntax instead of C++ syntax. For example, if you want to show the result type of the call `iface.activeLayer` type:

```
help(iface.activeLayer)
```

The following lines will be displayed:

```
Help on built-in function activeLayer:
activeLayer(...) method of qgis._gui.QgisInterface instance
activeLayer(self) -> QgsMapLayer
Returns a pointer to the active layer (layer selected in the legend)
```

This shows that the `activeLayer` call returns data that is of the `QgsMapLayer` data type.

The Python `dir()` function gives you more detailed information, showing a list of all the methods belonging to a class.

Try typing `dir(iface)` and compare it with the result of the previous `help(iface)` command.

Creating a plugin structure with Plugin Builder

A QGIS plugin can be created manually with a simple editor, but the simplest and most complete way to start to create a plugin is to use another Python plugin called Plugin Builder.

Plugin Builder generates the file infrastructure of the plugin, thus avoiding writing repetitive code. Plugin Builder creates only a basic and generic plugin, which can be modified to add specific user functionalities.

It will generate a generic plugin with the following GUI:

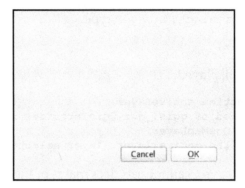

This is an almost empty dialog with two buttons. Every piece of this interface can be modified and customized to achieve the plugin's goal.

Installing Plugin Builder

The first step is to install Plugin Builder using the Plugin Manager by navigating to **Plugins | Manage and Install Plugins....**

Locating plugins

The **Plugin Builder**, as every third-party Python plugin is by default, is installed in your home directory at the following path:

```
<your homepath>/
AppData/Roaming/QGIS/QGIS3/profiles/default/python/plugins/
```

Over here, you'll find your Plugin Builder code at the `pluginbuilder` directory. You will notice that each installed plugin has a proper code directory. We'll create a new plugin that, to be loaded by default by QGIS, has to be created in the Python plugin directory. It's possible to change the default plugin directory path, but this is outside the scope of this topic.

Creating my first Python plugin - plugin_first

Starting the Plugin Builder will open a GUI to insert the basic parameters to set up the generation of your first QGIS plugin. The wizard-style interface is shown in the following screenshots:

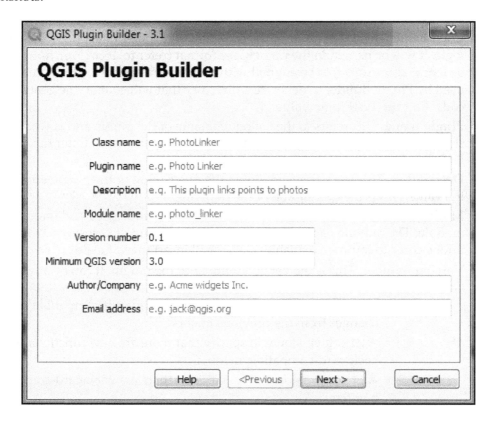

Each parameter is self-explanatory through tooltips which can be seen by moving the cursor over each parameter line.

In the fourth interface, it's possible to configure to generate plugin subdirectories filled with template code for internationlization, testing and so on. Every plugin should have tests and internationalization, but for the purpose of the tutorial, we can safely leave these options out with their related subdirectories.

Setting mandatory plugin parameters

There is a set of mandatory parameters that are always checked by the QGIS plugin repository when a plugin is uploaded. These parameters are also manually checked by QGIS members to approve the plugin officially in the central repository. The parameters are as follows:

- **Class name**: This is the name of the class that will contain the plugin business logic. It will be named in the CamelCase format (refer to `http://en.wikipedia.org/wiki/CamelCase`) to be aligned with the Python standard. This name will be used by Plugin Builder to generate a directory that will contain the generated code. Edit the TestPlugin value.

- **Plugin name**: This refers to the colloquial name of the plugin and is what will be shown in the Plugin Manager and in the QGIS Plugins menu. Enter the value My First Test Plugin.

- **Description**: This is a string containing the description or the plugin scope. Enter the value This is the description of the plugin.

- **Module name**: In Python, a group of classes can be addressed and imported as a module. The module name should be in lowercase and, if necessary, with underscores to improve readability. We shall insert the value `test_plugin`.

- **Version number**: This is the version number of the plugin. It can be any number. Generally, the versioning has this format, `<MAJOR>.<MINOR>.<PATCH>`, where:
  - A `MAJOR` version will specify that there are incompatible API changes from the previous majors
  - A `MINOR` version will specify that there are new functionalities in a backward-compatible manner
  - A `PATCH` version will specify that there are backward-compatible bug fixes
  - At the moment, we can leave the default value set at 0.1.

- **Minimum QGIS version**: This refers to the minimum QGIS version in which the plugin will run. Each QGIS version has its own API set; the plugin can be compatible with a specified newer version but not with older ones if it's not programmed to be compatible. The minimum QGIS version is used by the QGIS Plugin Manager to show only plugins that are compatible with the running QGIS. This means that in QGIS 2.0, it's not possible to see the plugin for 1.8 or plugins that are designed to work only with 2.4 or newer versions. We can leave the default value as 2.0.

- **Author/Company** and **Email address**: The parameters are obvious and are used to contact the developer if a user finds problems in the plugin. For example, you can set your name, surname, or company name, and your e-mail address.
- **About**: A detailed description of the plugin that will be shown in the Plugin Manager.
- **Menu**: In what menu have to be placed out plugin. We can leave the default value.
- **Bug tracker**: It's good practice to maintain a service to track the bugs of the plugin. Plugin users can file issues by preparing test cases that help to reproduce the bug. Tracking traces of the bugs and their solutions helps us to know the evolution of the plugin. Usually, the use of a VCS web service, as shown previously, provides a bug-tracking service. For example, the bug tracker for the plugin called Plugin Builder is provided by the QGIS infrastructure and can be found at `http://hub.qgis.org/projects/plugin-builder/issues`. For our plugin's purposes, we can write something meaningful.
- **Repository**: A repository can be added later; it is the location where the plugin code is located. Its common to use a **Version Control System** (**VCS**) repository to maintain your code. Some popular VCSes are **Git** or **Subversion**, and some others related to Git. There are famous online services available at `https://github.com/` and `https://bitbucket.org/product/`, where you can upload your project and maintain modifications. For example, the repository of the code of the Plugin Builder is `https://github.com/g-sherman/Qgis-Plugin-Builder`. For our plugin's purposes, we can write something meaningful.

Setting optional plugin parameters

There are also optional parameters that are really useful if your plugin will be made available for other users. The parameters are as follows:

- **Home page**: If the plugin has a web page where it is described, it's good practice to add a plugin home page where you can leave instructions on the usage of the plugin. We can leave this blank to start with.
- **Tags**: This field is really important for allowing QGIS users to find the plugin. It's used by the Plugin Manager to look for plugin keywords. For example, if the plugin is managing GPS data, its tags could be gps, gpx, satellite, and so on. Try to find the tags that best describe the plugin and edit them separated with commas.

- The last checkbox of the Plugin Builder interface is checked if the plugin is in the experimental stage. By default, the Plugin Manager shows only plugins that are not experimental. To list the experimental plugins, it's necessary to tick the relative checkbox option in the Plugin Manager configuration. During the first developmental stage of the plugin, it's good practice to set it as experimental.

Generating the plugin code

After setting all the necessary plugin parameters, it's time to generate the code by clicking on the last **Next** button, which will open a path selection dialog, which will select the location of the new plugin. Selecting the same directory that contains the plugin, Plugin Builder allows QGIS to find the new plugin. The default path should be:

```
<your homepath>/AppData/Roaming/QGIS/QGIS3/profiles/default/python/plugins/
```

However, you can create your plugin anywhere. Just remember to link or deploy it in the plugin directory to allow QGIS to load it.

After selecting a path, the code will be generated, creating a new directory in the selected path. In our case, the new directory will have the name `plugin_first`.

At the end of the code generation, a dialog will appear with a message explaining the steps to complete plugin creation.

The generated plugin is not available yet; it's necessary to restart QGIS and activate it in the Plugin Manager interface. The plugin is now fully functional, but after the first activation, its button in the QGIS toolbar will be without an icon.

Compiling the icon resource

To make the icon visible, it's necessary to compile the icon resource in order to have it available in Python. Resource compilation is a process to render an icon's platform independent of the Qt framework, which is the graphical infrastructure on which QGIS is built.

To compile the icon resource, it's necessary to use `pyrcc5` command. But before that, we need to run the batch file `qt5_env.bat` to make the Qt5 environment ready to use. Open the **OSGeo4W Shell**, and type the following and press enter.

```
qt5_env.bat
```

We also need to set the path for Python 3 environment just by typing:

```
py3_env.bat
```

We can see the results of py3_env.bat as following:

The pyrcc5 command is the Qt resource compiler for Qt5 and it's available in the pyqt5-dev-toolsqt5-designer packages.

Now, we need to go inside our new plugin plugin_first directory by using cd.

```
py3_env.bat
cd
C:\Users\User\AppData\Roaming\QGIS\QGIS3\profiles\default\python\plugins\plugin_first
```

As you can see here:

```
C:\Users\User>cd C:\Users\User\AppData\Roaming\QGIS\QGIS3\profiles\default\pytho
n\plugins\plugin_first

C:\Users\User\AppData\Roaming\QGIS\QGIS3\profiles\default\python\plugins\plugin_
first>
```

We can see in the second line that we are inside the plugin directory plugin_first now. Now, use pyrcc5 to compile the resources.qrc file, generating the Python version, resources_rc.py:

```
C:\Users\User\AppData\Roaming\QGIS\QGIS3\profiles\default\python\plugins\plugin_
first>pyrcc5 -o resources.py resources.qrc
```

Now restart QGIS and click **Plugins** | **Manage and Install Plugins...**

Now click **Settings** | **Show also experimental plugins** and and click on **All**, we will see our new plugin with an icon which is the default icon set by Plugin Builder.

To change the icon, just change `icon.png` for a new image, leaving the filename unchanged, and then recompile the icon resource.

 It's possible to change the filename and add more icon resources, but this is out the scope of the current chapter, so please refer to the Qt documentation for this.

Plugin file structure - where and what to customize

Our `plugin_first` code has been created in this folder:

```
<your homepath>/AppData/Roaming/QGIS/QGIS3/profiles/default/python
/plugins/
```

Here, we can find a complex file structure, where only a subset of files are strictly necessary for plugins and are in the scope of this book. The basic files are the following ones:

- `__init__.py`
- `metadata.txt`
- `Makefile`
- `icon.png`
- `resources.qrc`
- `resources.py`
- `plugin_first_dialog_base.ui`
- `plugin_first_dialog.py`

Each file has its own role inside the plugin, but only a few of them have to be modified to develop a custom plugin.

Other than basic files, the Plugin Builder can generates other files and directories useful for managing more complex plugin projects. The files and directories are as follows:

- `help/`
- `i18n/`
- `scripts/`
- `test/`

- `pylintrc`
- `plugin_upload.py`
- `README.html`
- `README.txt`

Exploring main plugin files

Here, we will describe the role of each of the main files that compose a plugin:

- The `__init__.py` file is the common Python module starting file and it's also the entry point for QGIS to load the plugin. Usually, it doesn't have to be modified to create a plugin

- The `metadata.txt` file is a text file containing all the information about the plugin. This file is read by Plugin Manager to manage the plugin inside QGIS. For example, in this file there are plugin classification tags or the minimum QGIS version in which the plugin can be run.

- `Makefile` is a set of instructions used by the `make` command to compile resources, manage some shortcuts to compile documentation, clean previously compiled files, and so on. Usually, it's not necessary to edit it.

- The `icon.png` file is the plugin icon. As explained previously, it will be modified with a definitive plugin icon.

- The `resources.qrc` file is the file that instructs Qt about how to manage the icon. Usually, it's not necessary to edit it, other than adding more icons or changing the filename of the `icon.png` file. More information about the resource file can be found in the Qt documentation at `https://doc.qt.io/qt-5.6/resources.html`.

- The `resource_rc.py` file is the compiled version of the `resource.qrc` file, and it's generated after compilation with the `pyrcc5` command.

- The `plugin_first_dialog_base.ui` file is a file in XML format, describing the layout of the user interface of the plugin. It's strictly necessary only if the plugin needs its own GUI. The GUI structure can be edited manually, but usually it's better to use the Qt framework to edit it. The framework is called **Qt Designer** or **Qt Creator** and it can be downloaded from `http://qt-project.org/downloads`. The GUI design with the Qt framework is beyond the scope of this book, but the framework has good tutorials explaining how to customize graphic interfaces.

- The `plugin_first_dialog.py` file contains the logic of the preceding plugin GUI layout. This is the place where you add the logic of the plugin related to the GUI, for example, buttons that are disabled when a specific value is inserted, and so on.

Plugin Builder-generated files

Plugin Builder can generate more than basic files, creating templates to manage complex Python plugin projects. A project can involve unit testing, detailed documentation, translation, code analysis, and so on. Here is a summary of these files:

- The help directory contains all the files necessary to automatically generate documentation in different formats, from HTML to PDF, for instance.
- The i18n directory contains files where we can add translations in other languages.
- The script directory contains some tools to facilitate the plugin development and deployment.
- The test directory contains unit tests for the plugins. It also contains utility classes to support unit testing.
- The `pylintrc` file is a configuration file for `Pylint`, a framework of code analysis.
- The `plugin_upload.py` file is a command-line utility to upload the plugin in the QGIS plugin repository.
- The `README` files contain the messages displayed at the end of plugin generation.

A simple plugin example

The goal of this section is to customize `plugin_first` to classify the loaded layers in the raster and vectors and respectively populate two combo boxes with the layer names.

Adding basic logic to TestPlugin

As said previously, to customize TestPlugin we have to modify some code portions in the files: `plugin_first_dialog_base.ui` for the GUI layout, `plugin_first_dialog.py` for the GUI logic, and `plugin_first.py` for the plugin logic.

Modifying the layout with Qt Designer

The default plugin GUI layout has only two buttons, **Ok** and **Cancel**. Here, we will add two combo boxes that will be populated by the logic of the plugin in `plugin_first.py`.

To edit the `plugin_first_dialog_base.ui` GUI layout, open it with Qt Designer, which will show the interface of the following screenshot:

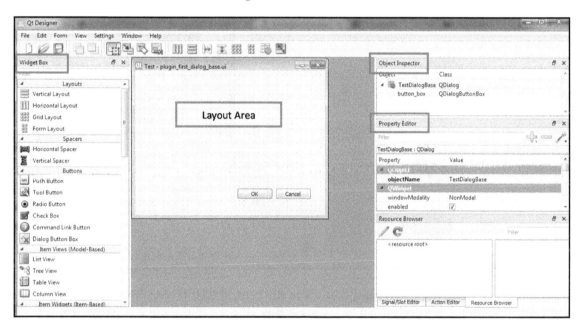

This is the graphical representation of the `plugin_first_dialog_base.ui` XML file. With Qt Designer we can reorganize the layout, adding new graphical elements, and also connect events and triggers related to the interface. In the preceding screenshot, the four gray boxes mark the designer sections:

- The **Layout Area** is the area where the plugin GUI is rendered
- The **Widget Box** section contains the list of predefined GUI components. Here, we'll look for the combo box to add to the GUI layout
- The **Object Inspector** section gives the hierarchy of graphical components composing the GUI layout
- The **Property Editor** section gives a list of all the properties of the graphical components that can be customized.

Adding two pull-down menus

The next steps will be to create the layout, as shown in the following screenshot:

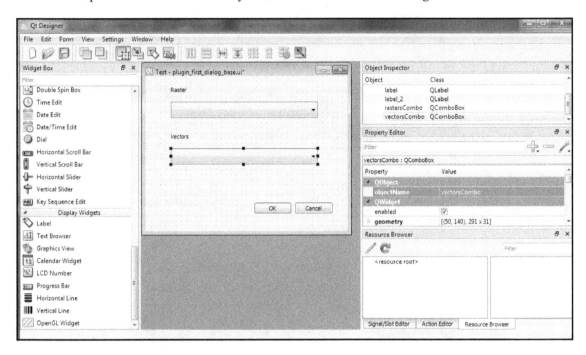

To create this layout, drag two combo boxes to the central GUI layout. The **Combo Box** option can be found by scrolling in the **Widget Box** section. In the same way, we'll add two labels on the top of each combo box.

To edit the labels, just double-click on them to enter label edit mode. This action is equivalent to changing the **Text** property in the **Property Editor** section.

After creating the layout, we can associate each combo box with an object name that will be used to distinguish the function of each combo box. To do this, we will change the **objectName** property in the **Property Editor** section. For example, we can set the name of the raster combo box as `rastersCombo`. In the same way, we rename the vector combo box as `vectorsCombo`.

The string used as **objectName** will be the name of the Python variable that refers to the graphical element. It will be used in the Python code when we want to get or set some property of the graphical element.

Modifying GUI logic

Our plugin doesn't need modification in the dialog code, `plugin_first_dialog.py`, because all the GUI updates will be guided by the `plugin_first.py` code directly populating the `rastersCombo` and `vectorsCombo` elements. This is a design decision; in other cases, it could be better or cleaner to have the logic inside the dialog code to hide combo names using the dedicated function added to the dialog.

Modifying plugin logic

Our core plugin logic has to be added. This can be done by modifying the `plugin_first.py` code. In many cases, such as in simple plugins or batch-processing instructions, only the `run(self)` function has to be modified inside the plugin code.

The `run()` function is the function that is called every time the plugin button is clicked on in the QGIS toolbox or the plugin is run from the plugin menu. In the `run()` function, we have to:

- Collect all loaded layers
- Classify in raster and vector layers
- Populate the combo boxes in the plugin GUI
- Show the GUI with the new values loaded

To get all the listed layers, we'll have to ask the container of all the displayed layers and project, that is, the `QgsProject` class. To do this, we will use the following code:

```
from qgis.core import QgsProject
layersDict = QgsProject.instance().mapLayers()
```

The first line imports the `QgsProject` class from the Python module `qgis.core`.

> You can import classes anywhere in the code, but it's good practice to import them at the beginning of the file.

Forgetting to import the class will cause an error during runtime. The error will be as follows:

```
Traceback (most recent call last):
File "<input>", line 1, in <module>
NameError: name 'QgsProject' is not defined
```

In the second line, the result variable, layersDict, is a Python dictionary; it contains a set of key-value pairs where the key is the unique ID of the layer inside QGIS, and the value is an instance of the QgsMapLayer class that can be a vector or raster layer.

> The variable, layersDict, has the suffix, Dict, only for a didactic reason, but it could simply be named layers.

Classifying layers

The next step is creating a list of vector and raster layer names. This can be achieved by looping layersDict in the following way:

```
from qgis.core import QgsMapLayer
vectors = []
rasters = []
for (id, map) in layersDict.items():
    if (map.type() == QgsMapLayer.VectorLayer):
        vectors.append( map.name() )
    elif (map.type() == QgsMapLayer.RasterLayer):
        rasters.append(map.name())
    else:
        print("Not Raster nor Vector for layer with id: " + str(id))
```

The first line, which is the import line, is necessary to allow the use of the QgsMapLayer class.

The next two lines initialize two empty Python lists that will be filled with the layer names. Adding an element to an array is done with the append Python command.

The layers are looped, separating each key-value pair directly into a couple of variables named id and map. The id variable is used to display a warning message in line 10.

Populating the combo box

After classifying the loaded layers, we set the values of the combo box of the plugin interface. This can be done with the following code:

```
self.dlg.rastersCombo.insertItems(0, rasters)
self.dlg.vectorsCombo.insertItems(0, vectors)
```

We named our combo boxes with object names, rastersCombo and vectorsCombo, in the two code lines earlier; combo boxes are populated with a standard QComboBox with the insertItems(...) call passing the two lists of layers. The 0 parameter is the index where we start to add new elements.

Understanding self

In the two preceding lines, there is a keyword, self, that may confuse everyone when approaching object-oriented programming for the first time. To explain it, try to follow where the TestPluginDialog() interface is created and saved.

At the beginning of the plugin_first.py code, the __init__(self, iface) function is where the Plugin GUI is created for the first time, with this instruction:

```
self.dlg = TestPluginDialog()
```

This means that the result of the creation of the dialog, the TestPlugiDialog() constructor call, is saved in the dlg variable that belongs to the current instance of the plugin. In this case, dlg is called an object or instance variable, where the instance of the plugin is referred to with the self variable.

The self variable is almost always available in every Python function; this allows us to access the dlg variable everywhere in the code.

Showing and running the dialog

We don't have to write any code here, because it will be generated by Plugin Builder. The action to show how a dialog is saved in the `self.dlg` instance variable is shown here:

```
# show the dialog
self.dlg.show()
# Run the dialog event loop
result = self.dlg.exec_()
# See if OK was pressed
if result:
    # Do something useful here - delete the line containing pass and
    # substitute with your code.
    pass
```

This code is self-explanatory with the comments generated by Plugin Builder.

Some improvements

As you can see for yourself, the plugin doesn't work well if it is run more than one time. The content of combo box grow on every run of the plugin. It's left to the reader to find a solution how to avoid this behavior.

More detail of the code

The complete code of this example, as usual, can be obtained from the source code for the book.

Here, we'll give a bird's-eye view of the `test_plugin.py` source code that contains the `TestPlugin` class. This class has other methods than `run(self)`; these are as follows:

- `__init__(self, iface)`
- `tr(self, message)`
- `add_action(self, icon_path, text, callback, enabled_flag=True, add_to_menu=True, add_to_toolbar=True, status_tip=None, whats_this=None, parent=None)`
- `initGui(self)`
- `unload(self)`

The following list provides a brief description of these methods:

- The `__init__(self, iface)` method is always present in every Python class and it is the constructor, which means that it is called every time you find a call, such as `TestPlugin(iface)`, as you can find in the `__init__.py` code.
- In our case, the constructor needs the `iface` variable passed as a parameter during construction.
- The constructor has the role of initializing the current translation, creating the plugin dialog GUI, and also creating the toolbar where the plugin button is to be added.
- The `tr(self, message)` method is just a shortcut to access the Qt translation engine of string messages.
- The `initGui(self)` method gets the icon resource and instructs how to interact with the QGIS menu, calling the `add_action(...)` method. This method is always called when the plugin is loaded in QGIS.
- It is important to remember the difference between loading a plugin and running a plugin.
- Loading is done using the Plugin Manager, or automatically at the start of QGIS if the plugin was already loaded in the previous session.
- Running the plugin is when the user starts the plugin by clicking on the plugin icon or activating it in the **plugin** menu.
- The `add_action(...)` method has a lot of parameters that allow for fine configuration, but most of them are used with their default values, `True` or `None`. The main goal of this method is to create the menu in the plugin menu and to create the button to call the `run()` method. In Qt, these kind of buttons are objects of the `QAction` class.
- The `unload(self)` method is used to unroll all the QGIS GUI elements added with the previous `add_action(...)` method.
- This method is always called when the plugin is unloaded in QGIS using the Plugin Manager.

Setting up a debugging environment

Software development is a complex task and there's no software without bugs. Debugging is the process to remove software failures. Debugging is a task that can involve some other software to facilitate the debugging process.

A plugin can become complex, requiring debugging tools to discover problems. The complexity of the debugging process can start by inserting some prints inside the code or adding log messages to finish controlling the execution instructions by instructing how to find execution problems.

Inserting a breakpoint to stop the execution at a certain point in the code of a third-party QGIS plugin can be useful for discovering how it works.

What is a debugger?

There is a set of possible tools to debug the Python code, but we'll focus only on **PyDev**, which reduces the number of installation steps and allows remote debugging without modification of the plugin code.

PyDev is an Eclipse plugin, where Eclipse is a free software programmable framework used to develop almost everything. PyDev can be added to a local installation of Eclipse, adding it from the marketplace, but to reduce the number of installation steps, it's suggested that you install Aptana Studio 3, an Eclipse customization with PyDev already installed.

Installing Aptana

First, we need to install Java SE Development Kit 8 from this site. Select **Java SE Development Kit 8u201** or higher and select the version according to the processor of your computer or your operating system.

For our case, download the .exe file corresponding to Windows x86 to make it compatible with Aptana Studio and execute the .exe file. Click on **3.7.1....** and execute the executable file.

The installation version of Aptana Studio 3 should be at least version 3.6.1, because some previous versions have bugs that don't facilitate code writing. If a version greater than 3.6 is not available, then it's necessary to upgrade to the beta version. To upgrade, follow the instructions mentioned at http://preview.appcelerator.com/aptana/studio3/ standalone/update/beta/. After that, download Aptana Studio from this site https:// github.com/aptana/studio3/releases. Similarly, download the version according to your operating system, for example, Aptana_Studio_3_Setup.exe for **Windows 64**.

Now, copy the **.exe** file to **C:** drive and using command prompt (in windows, click on the start menu or press the windows button and search using cmd), we can install this by first going to C drive inside command prompt, by writing cd \ and pressing enter and to install Aptana, use passive and norestart option along with the .exe file name to install without any interruption:

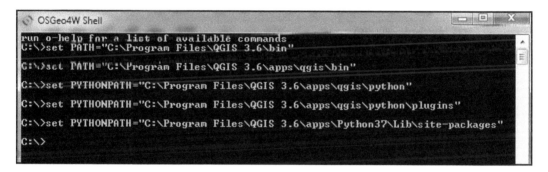

Now, we need to set the Path for Python by adding the path of Python in the Environment Variables. We can do this by typing the following commands in **OSGeo4W Shell**:

Please have a look into this page at stackoverflow for guide https://stackoverflow.com/questions/3701646/how-to-add-to-the-pythonpath-in-windows.

But, if you find adding the path of Python in the Environment Variables overwhelming, you can install **Python 3** from this address `https://www.python.org/downloads/release/python-370/`. After opening the .exe file by double clicking and tick besides **Add Python 3.7 to Path** and click on **Install Now** and then keep on accepting all the defaults and finish installation:

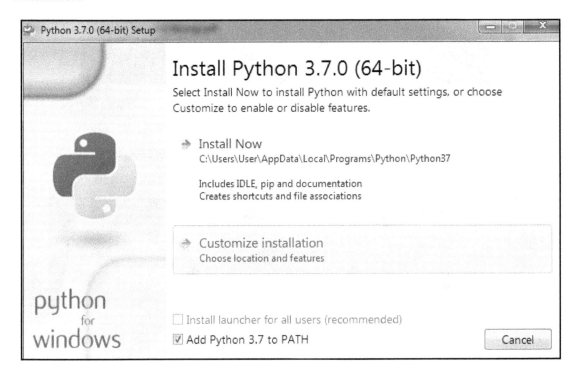

Now that you have added **Python** to **Environment Variables**, you will also need to install Pydevd server inside Aptana Studio and it can be done by clicking on **Help** | **Install**. We will get a new prompt and paste `http://www.pydev.org/updates` in the text box corresponding to **Work with:**. Select **PyDev for Eclipse** and then click on **Next** and then **Finish.** Only keep the option **Show only the latest versions of available software**. These steps will look something similar to the below image:

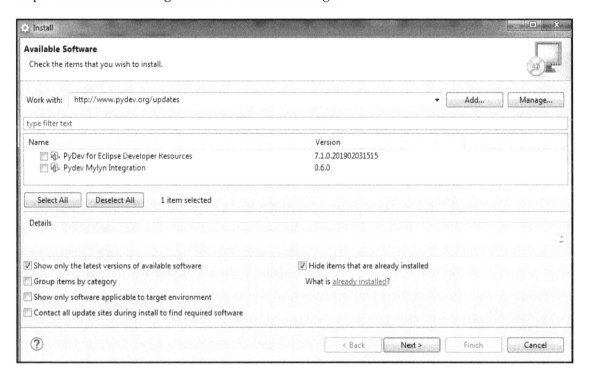

Now set the Python interpreter inside Aptana Studio by clicking **Windows | Preference | PyDev | Interpreters | Python Interpreter**. Now, click on **Browse for python/pypy exe** and browse to `<your homepath>\AppData\Local\Programs\Python\Python37-32\python.exe`. Then click on Apply and Close:

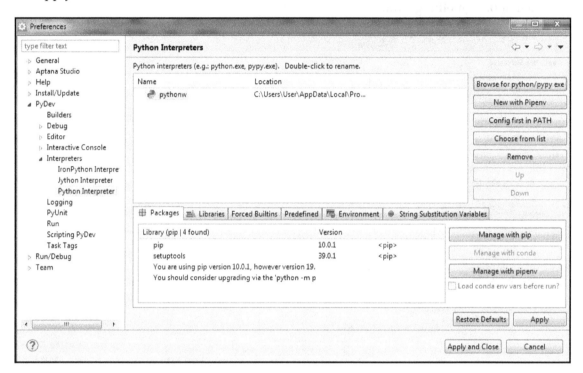

Setting up PYTHONPATH

To allow QGIS to be connected with the PyDev daemon, it's necessary that the PyDev daemon path be added to the `PYTHONPATH` environment variable.

To find the path to add, look for the `pydevd.py` file; it will be in the AptanaStudio installation path. If you find more than one version, get the path that has the highest version number.

For example, in my Windows installation, I have the following path to `pydevd.py` file and we will also use this path adjusting for our home directory (you might also need to change depending upon the operating system and version of the software you are using):

```
C:/Users/User/AppData/Roaming/Appcelerator/Aptana
Studio/plugins/org.python.pydev.core_7.1.0.201902031515/pysrc
```

This path can be added in the `PYTHONPATH` session, or directly in the QGIS environment by modifying `PYTHONPATH` by navigating to Setting | Options | System in the Environment section, as shown in the following screenshot:

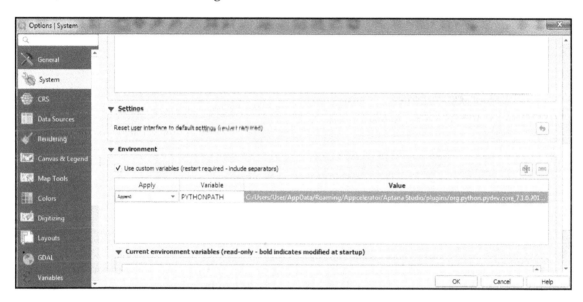

Now, close QGIS and then reopen it and now it will be possible to import the pydevd Python module into the QGIS Python Console and set the connection to the PyDev debug server, as described in the next section.

To test whether the path is set correctly, try to type the following code in the QGIS Python Console:

```
import pydevd
```

If it generates an error, it means that `PYTHONPATH` is not set correctly with the path of the `pydevd` module.

Starting the Pydevd server

The first step to connect to PyDev server is to start the server in the Aptana environment. This can be achieved by opening the Debug perspective of Aptana and then starting the server with the relative start/stop buttons. The buttons and Debug perspective are shown here:

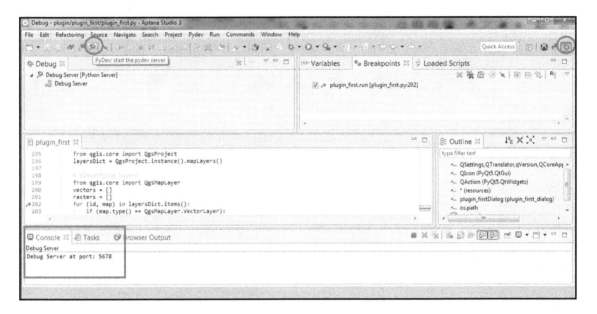

The Debug perspective can be opened by clicking on the right-hand side button circled in red. If the Debug perspective button is not available, it can be added to the Aptana menu. Navigate to **Window** | **Open Perspective** | **Other**.

The start/stop server button is shown by the red circle on the left.

While starting the server, some messages will appear in the **Debug** window highlighted by the red box in the upper-left section; this means that the server is running. In the **Debug** window, all connected clients will be listed.

In the red box at the bottom, the Aptana console window shows that the server answers to the port, 5678; this is the information that we'll use to connect from QGIS.

Connecting QGIS to the Pydevd server

After running the PyDev server, we'll connect to it from QGIS. In the QGIS Python Console, type the following code:

```
import pydevd
try:
    pydevd.settrace(port=5678, suspend=False)
except:
    pass
```

The preceding code first imports the `pydevd` module and then connects to the server with the `settrace` method. The connection is inside try/catch to allow catching an exception raised in case `settrace` cannot connect. The connection will take some seconds to connect. If the connection fails, the Python Console will show a message similar to the following one:

```
Could not connect to 127.0.0.1: 5678
Traceback (most recent call last):
File
"/mnt/data/PROGRAMMING/IDE/Aptana_Studio_3/plugins/org.python.pydev_3.8.0.2
01409251235/pysrc/pydevd_comm.py", line 484, in StartClient
s.connect((host, port))
File "/usr/lib/python2.7/socket.py", line 224, in meth
return getattr(self._sock,name)(*args)
error: [Errno 111] Connection refused
```

If the connection is successful, the Aptana Debug perspective will change, showing the connected clients.

Connecting using the Remote Debug QGIS plugin

The simplest way to set up PYTHONPATH and connect to the PyDev server is using the **Remote Debugger** experimental plug, which can be installed as usual. The main interface is shown in the following screenshot:

The **Remote Debugger** plugin is in charge of setting up PYTHONPATH writing it in the **pydevd path** textbox.

Debugging session example

Here, we will show how to debug the TestPlugin remotely. We'll also learn how to insert a code breakpoint, stop executions, and show variable values during executions. The steps to follow are:

1. Create a PyDev project that points to the source code of TestPlugin
2. Add a breakpoint to the TestPlugin run() function in the Aptana Debug perspective
3. Start the PyDev Debug server
4. Connect to the PyDev server from QGIS

5. Run the plugin
6. Explore the variable values
7. Continue the execution of the plugin

Creating a PyDev project for TestPlugin

To be able to add code breakpoints, it's necessary to load `plugin_first.py`. This can be simply opened as a file, but it's better to learn how to view the entire plugin as a PyDev project. This allows us to use Aptana as a debug and develop environment. This is done in two steps:

1. Create a PyDev project in Aptana Studio 3
2. Link the source code to the project

Creating a PyDev project called `plugin_first` is done by navigating to **File | New Project**. This will open a wizard where we'll have to look for a **PyDev Project** entry. Select it and click on the **Next** button at the bottom. Here, the wizard will pass to the phase to insert the project name, `plugin`, and then click on the **Finish** button at the bottom. A new project called `plugin` will be shown in the **PyDev Package Explorer** Aptana section.

The next step is to add the folder; right- of our `plugin` code inside the project. To add it, select the `plugin PyDev` project click on it to add a new folder, as shown in this following screenshot:

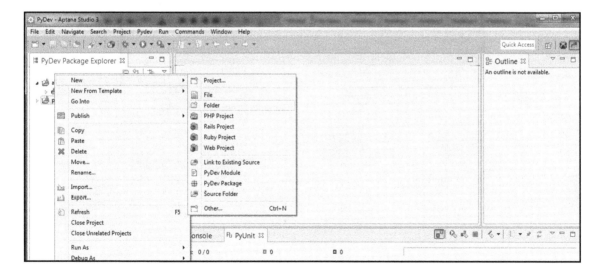

Here, we can see the contextual menu to add a new folder for the selected project. This action will open a GUI where we can create or link a new folder. In our case, it's useful to link to the existing plugin code, which can be done using the **Advanced** features of the GUI, as shown here:

After linking the folder, it will appear under the `TestPlugin` PyDev project, where we can look for the `test_plugin.py` code. Double-click on the file; it will be opened on the right-hand side of Aptana, as shown in the following screenshot:

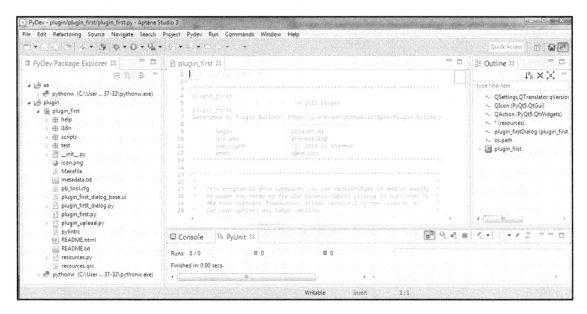

Adding breakpoints

Breakpoints are debugger instructions to stop execution at a specified line to allow users to investigate variable values and eventually change their values manually.

Our scope is to add a simple breakpoint and check that the plugin execution stops exactly at that point, passing control to the remote debugger.

To add a breakpoint, open the Debug perspective and double-click on the left-hand side of the line number, for example, the line with the `for (id, map)` code in `layersDict.items()`. Aptana will add a breakpoint, as shown in this screenshot:

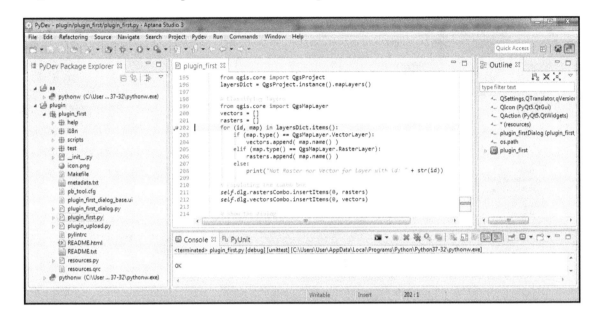

The line where the breakpoint is added is marked by the red box on the left, and a new breakpoint will be listed in the **Breakpoints** list, as marked by the red box in the upper-right corner.

Debugging in action

Now, it's time to test the debug session. Let's start the PyDev debug server and connect to it, as described previously.

In QGIS, run `TestPlugin`; QGIS will now freeze because it's starting to execute the run() method of the `plugin_first.py` code and a breakpoint is encountered. So, the control is passed to the PyDev debugger.

A detailed description of how to work with the `PyDev` Debug perspective is out of the scope of this chapter, but it's possible to find more documentation on PyDev at `http://pydev.org/manual_adv_debugger.html`.

Summary

In this chapter, we approached three important topics on developing plugins: first, how to get help to solve programming problems; second, how to create a basic plugin as a template to develop more complex plugins; and finally, how to debug it. These topics demonstrate basic skills for managing plugin development, which can become complex during design and development.

The chapter focused on creating a basic infrastructure to easily solve problems that could be found during the development of a working plugin.

In the next chapter, we will explore PyQGIS programming in depth and learn how to manage raster, vector, algorithm, and QGIS interface interactions. These skills will be useful for adding specific business functions to QGIS to solve practical processing problems.

11
PyQGIS Scripting

This chapter is focused on a specific use case or user. QGIS can be used in many different ways, and the GIS user is an eclectic user who has many different ways to interact with data and QGIS tools. The main focus of this chapter is a user who has an algorithm and wants to integrate it with QGIS.

An algorithm can be an external program, such as a water-modeling tool, or a Processing Toolbox's set of instructions.

It's possible to interact with QGIS in many different ways, from experimenting with PyQGIS in the Python console to creating plugins that control events generated by QGIS. This chapter will give you an overview of the following topics:

- Where to learn Python basics
- How to load layers
- Vector structure
- Iterating over features
- Editing features
- Running Processing Toolbox algorithms
- Interacting with map canvas

Where to learn Python basics

This chapter is not intended to give you an introduction to Python programming. There are a lot of free online resources and **Massive Open Online Course** (**MOOC**) (http://en.wikipedia.org/wiki/Massive_open_online_course) courses on the web.

The main resources can be obtained directly from the Python home page at https://www.python.org/about/gettingstarted/, where there is a big collection of guides, free books, and tutorials.

Tabs or spaces – make your choice!

When programming in Python, it's important to give special attention to editing code with correct indentation. Avoid mixing spaces and tabs, because it can generate errors that can be difficult to understand, especially for someone who is a beginner at Python programming. The official PEP 8 Python standard recommends using spaces.

How to load layers

Loading layers in QGIS involves two different steps, which are as follows:

1. Load the layer. This step creates a variable with the layer information and related data.
2. Register the layer in QGIS so that it can be used by other QGIS tools.

Loading a layer means loading a reference to the layer and its metadata. The layer is not necessarily loaded in memory, but is usually fetched only when data is accessed to be processed or visualized.

Loading and registering a layer are separate steps. A layer can be loaded, processed, and modified before it is visualized, or it can be loaded as temporary data for an algorithm. In this case, it's not necessary for the QGIS framework to be aware of the layer.

The `iface` object has shortcuts to load raster and vector layers in a single step instead of loading and registering them in separated steps.

Every layer type is managed by a provider manager. QGIS has some internal implemented providers, but most of them are external libraries. The list of available providers depends on the QGIS installation. This list can be obtained by typing the following code snippet in the QGIS Python console:

```
>>> QgsProviderRegistry.instance().providerList()
['DB2', 'WFS', 'arcgisfeatureserver', 'arcgismapserver', 'delimitedtext',
'gdal', 'geonode', 'gpx', 'mdal', 'memory', 'mesh_memory', 'mssql', 'ogr',
'oracle', 'ows', 'postgres', 'spatialite', 'virtual', 'wcs', 'wms']
```

The preceding result shows a Python list of strings that have to be used when a PyQGIS command needs the provider parameter.

How to manage rasters

Like most free software projects, the QGIS community doesn't want to reinvent the wheel if it's not strictly necessary. For this reason, most raster formats that are managed by QGIS can be loaded, thanks to the GDAL library, which is documented at `http://gdal.org/`.

In this chapter, we will use sample data from QGIS that can either by clicking an option to install the QGIS sample dataset while using Windows installer, or this can be downloaded from this link `https://qgis.org/downloads/data/qgis_sample_data.zip`. To code the loading of our first raster, named `landcover.img`, which is available in the `qgis_sample_data` folder, execute the following code snippet in the QGIS Python console by adapting the path to `landcover.img` based on your operating system and data location:

```
myRaster =
  QgsRasterLayer("/qgis_sample_data/raster/landcover.img")
```

In this way, the layer is loaded and referred to with the `myRaster` variable. If we want the layer to be visible in the legend with the name `MyFirstRaster`, we need to modify the preceding code snippet by adding a second parameter, as follows:

```
myRaster =
  QgsRasterLayer("/qgis_sample_data/raster/landcover.img",
  "MyFirstRaster")
```

 Two things should be noted: the first is that loading a raster layer is usually not necessary to specify the raster provider because it is GDAL by default. The second is that loading a layer is not the same as visualizing it in QGIS; a layer reference is loaded in memory to be processed and it is eventually visualized.

One of the basic actions after loading a layer, raster, or vector, is to ensure that it has been loaded correctly. To verify this, execute the following code:

```
myRaster.isValid()
```

It should return `True` if the layer has been loaded correctly. The following snippet performs some recovery actions if loading fails:

```
if not myRaster.isValid():
<do something if loading failed>
```

Exploring QgsRasterLayer

Any raster is stored in an object of the `QgsRasterLayer` class, so it's important to explore the methods of this class. In the preceding code, the `myRaster` variable is an instance of the `QgsRasterLayer` class. This means that all methods of the raster are documented at `http://qgis.org/api/classQgsRasterLayer.html`. This class is a specialization of the generic `QgsMapLayer` class.

Most Python QGIS classes are direct bindings to their C++ version. This means that API documentation refers only to C++ APIs, but not all methods are visible to Python. If you want to have all the methods available in Python, use the Python help command by typing the `help(QgsRasterLayer)` command in the Python console.

For example, we can get some raster information by calling the methods, as follows:

```
print(myRaster.height(), '-', myRaster.width())
```

The preceding code will produce the following output:

```
1964 - 3663
```

To get the extent of the layer, it is necessary to use the `extent()` method of the `QgsMapLayer` class; so, execute the following code:

```
print(myRaster.extent())
```

This will generate a strange result that is similar to the following output:

```
<QgsRectangle: -7117600 1367760, 4897040 7809680>
```

This shows that the result of the `extent()` method is a `QgsRectangle` instance where it is possible to call all the methods belonging to `QgsRectangle`. For example, the bounding box coordinates can be printed with the following code snippet:

```
ext = myRaster.extent()
print(ext.xMinimum(), ext.yMinimum(), '-', ext.xMaximum(), ext.yMaximum())
```

This will produce the following result:

```
-7117600.0 1367760.0 - 4897040.0 7809680.0
```

Visualizing the layer

Finally, we can visualize the raster using the centralized QGIS layer manager, called QgsProject. This class is like the hub where we can manage layer loading and unloading. It's useful to read the list of its methods in the QGIS API documentation.

QgsProject is a singleton class. This means that it can't be instantiated multiple times like QgsRasterLayer. For example, we can have different loaded raster layers, and each one is an instance of the QgsRasterLayer class, but it's not possible to have different QgsProject instances. This is because it is blocked by code and it's possible to get only the unique instance using the instance() method.

Finally, to visualize the layer, we have to execute the following code:

```
QgsProject.instance().addMapLayer(myRaster)
```

This will produce an output similar to the following:

```
<qgis._core.QgsRasterLayer object at 0x0000000008EB55E8>
```

After the image is loaded, QGIS will appear as shown in the following screenshot:

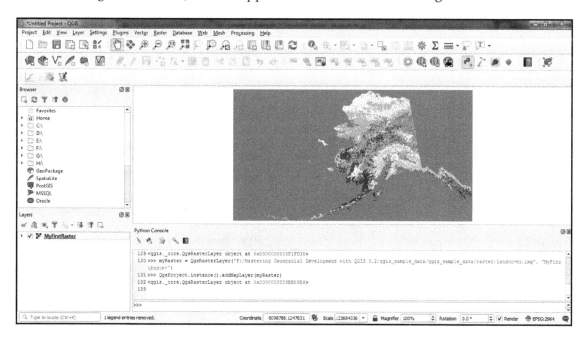

The preceding screenshot displays the image loaded in QGIS with a default false-color rendering palette.

Managing vector files

Similar to the raster layer in the previous section, most of the vector formats managed by QGIS are supported by the OGR library, a part of the GDAL library. OGR is documented at the same link of the GDAL library. All vector formats managed by OGR are listed at `http:/ /www.gdal.org/ogr_formats.html`.

To read vector data, it's always necessary to specify the provider because it can be provided by different sources.

To code the loading of our first shapefile, named `alaska.shp`, which is available in `qgis_sample_data`, execute the following code snippet in the QGIS Python console:

```
myVector =
  QgsVectorLayer("/qgis_sample_data/shapefiles/alaska.shp",
  "MyFirstVector", "ogr")
```

This way, the layer is loaded and referred to by the `myVector` variable. After adding the layer to `QgsMapLayerRegistry`, it will be visualized in the legend with the name `MyFirstVector`. In the `QgsVectorLayer` constructor call, it's possible to find a third string parameter, `ogr`, which specifies that the OGR library should be used to load the `alaska.shp` file.

As usual, we will check whether the loaded layer is valid using the following code:

```
myVector.isValid()
```

Managing database vectors

If vector data is hosted in a spatial database, it can be loaded by specifying the location and connection information using the `QgsDataSourceURI` class. A **Uniform Resource Identifier (URI)** is how a resource can be identified on a network such as the World Wide Web.

The following code snippet shows how to fill the URI with the necessary information to connect to a remote spatial database such as PostGIS (`http://postgis.net/`):

```
uri = QgsDataSourceURI()
uri.setConnection("localhost", "5432", "myDb", "myUserName",
  "myPassword")
uri.setDataSource("public", "myTable", "the_geom", "myWhere")
print(uri.uri())
```

The first line creates an instance of `QgsDataSourceURI` that is filled with other information in the next lines.

The `setConnection` method accepts the IP or the symbolic name of the database server, the connection port, the database name, the username, and the password. If the password is set to `None`, QGIS will ask you for the password for connecting to the database.

The `setDataSource` parameter refers to the schema name, the table name, and the geometry column where the geometry is archived. Finally, an optional `where` string could be set to directly filter data, in this case, `myWhere`.

> You can load a query without having a `Db` table counterpart. Just write your query instead of the table name, and place round brackets around it. For example, the previous `setDataSource` method will become `uri.setDataSource("public", "( <here your query> )", "the_geom", "myWhere")`.

The last line shows you the URI string that will be used to point to the vector data. It will be a string similar to the following one:

```
dbname='myDb' host=localhost port=5432 user='myUserName'
  password='myPassword' table="public"."myTable" (the_geom)
  sql=myWhere
```

An alternative way is to create the URI string manually, rather than populating the `QgsDataSourceURI` class, but it's generally more readable and less error-prone to write the previous code than a complex string.

If the database is on a SpatiaLite (`http://www.gaia-gis.it/gaia-sins/`) file, it's necessary to substitute the `setConnection` method with the following code:

```
uri = QgsDataSourceURI()
uri.setDatabase("/path/to/myDb.sqlite")
uri.setDataSource("", "myTable", "the_geom", "myWhere")
print(uri.uri())
```

This generates the following URI string:

```
dbname='/path/to/myDb.sqlite' table="myTable" (the_geom)
  sql=myWhere
```

After the URI string is created, we can use it to create a new vector layer with the following code:

```
myVector = QgsVectorLayer(uri.uri(), "myVector", "postgres")
```

The third string parameter specifies the data provider, which, in the case of vector data that is hosted on a SpatiaLite database, would have the `spatialite` value.

As usual, to visualize the vector, we have to use the following code:

```
if myVector.isValid():
    QgsProject.instance().addMapLayer(myVector)
```

The preceding code visualizes the vector only if it has been correctly loaded. Failures can happen for reasons such as errors in parameter settings, a restriction of the vector provider, or a limitation by the database server. For example, a PostgreSQL database could be configured that would allow access to a vector table only for a specific group of users.

Vector structure

To describe the `QgsVectorLayer` class, we will first approach basic layer parameters, and then we will explore how the vector is organized. We will explore some classes that are involved in the vector structure that represent rows and headers.

The basic vector methods

We will explore the vector class by working on a real vector; we will load `alaska.shp` in the `myVector` variable.

This variable is an instance of the `QgsVectorLayer` class. This means that all methods of the vector are documented at `http://qgis.org/api/classQgsVectorLayer.html`. As for rasters, this class is a specialization of the generic `QgsMapLayer` class.

To get the extent of the layer, it's necessary to use the `extent()` method of the `QgsMapLayer` class:

```
print(myVector.extent().toString())
```

Executing the preceding code will generate the following result:

```
-7115212.9837922714650631,1368239.6063178631011397 :
4895579.8114661639556289,7805331.2230994049459696
```

This shows the corner coordinates in the format of `xmin,ymin: xmax,ymax`.

To know how many records or features contain the vector, use the following code:

```
myVector.featureCount()
```

This will produce a result of `653` records.

As we saw earlier, vectors can be sourced from different providers, each one with its proper capabilities and limitations. To discover the capability of `myVector`, use the following method:

```
myvector.capabilitiesString()
```

This will give the following result:

```
'Add Features, Delete Features, Change Attribute Values, Add Attributes,
Delete Attributes, Rename Attributes, Create Spatial Index, Create
Attribute Indexes, Fast Access to Features at ID, Change Geometries'
```

This describes all the possible actions available on the vector.

Describing the vector structure

Compared to rasters, vectors are more complex. A vector involves a set of classes that are used to represent every piece of the vector, from the header to a single attribute.

We can think of a vector as a table with rows and columns, and one header that describes each column of the table. Each row has its own geometry and attributes that are archived in the columns of the row. Each vector can only contain a geometry type. For example, they can only be composed of points, lines, polygons, or collections of these geometry types.

The structure of the classes involved in a vector table is shown in the following diagram:

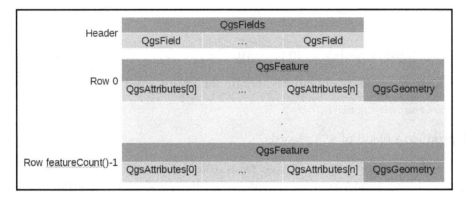

A description of these classes is the subject of the following paragraphs.

Describing the header

The container of the header information is the `fields` class, which contains methods to point to every column description. A column description is abstracted by the `fields` class.

To get the header container for `myVector`, we use the following code:

```
header = myVector.fields()
```

Here, `header` will store a `fields` instance. We can use `fields` methods to add or remove columns, get the column index by name with `indexFromName`, or get a specific field using its index with the following code:

```
field_0 = header[0]
```

To iterate on all column fields, use this snippet:

```
for column in header:
<do something with the variable "column" that is a field>
```

Each field has its methods to obtain the column name with `name()`, get its `type()` and `typeName()` methods, and also get its `precision()` method if it is numeric.

In the following code, we can explore the characteristics of the header of `myVector`:

```
header = myVector.fields()
print("How many columns?", header.count())
print("does the column 4 exist?", header.exists(4))
```

```
print("does the column named 'value' exist?",
header.indexFromName('value'))
print("Column 0 has name", header[0].name())
for column in header:
    print("name", column.name())
    print("type", column.typeName())
    print("precision", column.precision())
```

This will produce the following output:

```
How many columns? 3
does the column 4 exist? False
does the column named 'value' exist? -1
Column 0 has name cat
name cat
type Real
precision 0
name NAME
type String
precision 0
name AREA_MI
type Real
precision 15
```

Describing the rows

Each row of `myVector` is a `QgsFeature` instance. The feature contains all attribute values and a geometry. There are also as many values as in `QgsField` in the header.

Each feature has its own unique ID, which is useful for retrieving the feature directly. For example, to retrieve the feature with an ID of 3, we use the following code:

```
features = myVector.getFeatures(QgsFeatureRequest(3))
feature = next(features)
```

The preceding code could appear overcomplicated to retrieve a single feature; in the next section, we will show you the reason for this complexity by explaining the role of the `QgsFeatureRequest` class in retrieving features.

With this feature, we can retrieve all the attributes and the geometry of the feature using the following code:

```
print(feature.attributes())
geom = feature.geometry()
print(geom)
```

This will produce something similar to the following result:

```
[4, 'Alaska', 0.322511]
<QgsGeometry: MultiPolygon (((-194626.39810467543429695
7664103.45701420214027166, -196616.8670895871182438
7659215.00832086149603128, -196562.23079375541419722
7668383.29036159440875053, -194626.39810467543429695
7664103.45701420214027166)))>
```

This shows an array of values we get from the `attributes()` call and an instance of a `QgsGeometry` result of the `geometry()` method that is saved in the `geom` variable.

Exploring QgsGeometry

The `QgsGeometry` class is a complex and powerful class that can be used for a lot of geometry operations, and most of them are based on the capability of the underlying GEOS library.

Geometry Engine Open Source (GEOS) is an extensively used library for managing geometry entities. You can find more information about this at http://trac.osgeo.org/geos/.

It's difficult to describe the richness of this class and all the available methods; for this reason, it's best to read the API documentation. Here, we will give you only a brief introduction to some of the useful and commonly used methods.

It's possible to have the `length()` and `area()` methods of the geometry, when these values have sense depending on the geometry `type()`. For the `geom` variable that we got in the previous paragraph, the following code prints its length, area and whether it is multipart geometry or not:

```
print("it's length is", geom.length())
print("it's area measure", geom.area())
print("Is it multipart?", geom.isMultipart())
```

This will generate the following output:

```
it's length is 19143.8757902
it's area measure 8991047.15902
Is it multipart? False
```

The area and length unit depend on the `myVector` CRS and can be obtained with the following code:

```
myVector.crs().mapUnits()
```

This returns 0 for `QGis.Meters`, 1 for `QGis.Feet`, and 2 for `QGis.Degrees`.

Other interesting methods of the `QgsGeometry` class are related to spatial operators such as `intersects`, `contains`, `disjoint`, `touches`, `overlaps`, `simplify`, and so on.

Useful methods can be found to export geometry as a **Well-Known Text** (**WKT**) string or to GeoJSON with `exportToWkt` and `exportToGeoJSON`.

There are a bunch of static methods that can be used to create geometry from WKT or QGIS primitives such as `QgsPoint` and `QgsPolygon`. A static method is a method that can be called without an instance variable. For example, to create geometry from a point that is expressed as a WKT, we can use the following code:

```
myPoint = QgsGeometry.fromWkt('POINT(-195935.165 7663900.585)')
```

Notice that the `QgsGeometry` class name is used to call the `fromWkt` method. This is because `fromWkt` is a static method.

In the same way, it's possible to create `myPoint` from `QgsPoint` with the following code:

```
newPoint = QgsPointXY(-195935.165, 7663900.585)
myPoint = QgsGeometry.fromPointXY(newPoint)
```

The two methods that we just described are equivalent, and generate `QgsGeometry` in the `myPoint` variable.

Iterating over features

Now, it's time to discover how to get all the features or a subset of them. The main way to iterate over all features or records of `myVector` is by using the following code, which shows the ID of each feature:

```
for feature in myVector.getFeatures():
    feature.id()
```

This will print a list of all 653 record IDs, as shown here:

```
0
1
...[cut]...
652
```

It's not always necessary to parse all records to get a subset of them. In this case, we have to set the `QgsFeatureRequest` class parameters to instruct `getFeatures` and then retrieve only a subset of records; in some cases, we must also retrieve a subset of columns.

The following code will get a subset of features and columns:

```
rect = QgsRectangle(1223070.695, 2293653.357, 9046974.211,
 4184988.662)
myVector.setSubsetString('"AREA_MI" > 1000')
request = QgsFeatureRequest()
request.setSubsetOfAttributes([0, 2])
request.setFilterRect(rect)
for index, feature in enumerate(myVector.getFeatures(request)):
    print("The record %d has ID %d" % (index, feature.id()))
```

This will produce the following list of eight records:

```
The record 0 has ID 223
The record 1 has ID 593
The record 2 has ID 596
The record 3 has ID 599
The record 4 has ID 626
The record 5 has ID 627
The record 6 has ID 630
The record 7 has ID 636
```

In the preceding code, the first line creates a `QgsRectangle` method that is used in the `setFilterRect()` method to get only features that are within the rectangle. Then, only the values of columns 0 and 2 are fetched, setting the filter with `setSubsetOfAttributes`.

It should be noted that the `QgsFeatureRequest` class has the `setFilterExpression` method, which is useful for selecting features that are bigger than 1,000, but it can't be used in the previous case. The `QgsFeatureRequest` code forces the exclusive use of a spatial filter or an expression filter. For this reason, to filter by expression and by bound box at the same time, it's necessary to set the expression on a layer level with the `setSubsetString` method.

The `enumerate` statement is a Python instruction to get something that can enumerate and return pairs of elements, with the first element as the index and the second as the enumerated element.

Describing iterators

The preceding code uses the PyQGIS `getFeatures` statement that doesn't return features directly, but instead returns an iterator. In fact, executing the `myVector.getFeatures()` code produces an output similar to the following:

```
<qgis._core.QgsFeatureIterator object at 0xd023c20>
```

An iterator is a Python object that gets a record every time it is asked to get one; in our case, the `for` statement requests a record after every iteration.

The iterator works like a proxy: it doesn't load all the features in memory, but gets them only when it is necessary. In this way, it is possible to manage big vectors without memory limitations.

Editing features

After being able to parse all the features, it's necessary to learn how to modify them to satisfy our processing needs. Features can be modified in two ways:

- Using the data providers of the vector
- Using the methods of `QgsVectorLayer`

The difference between these two methods is the ability to interact with some editing features of the QGIS framework.

Updating the canvas and symbology

We will now modify the Alaska shapefile in the following subsections. If we modify some geometry of the legend classification, it will be necessary to refresh the canvas and/or layer symbology. The canvas can be refreshed with the following commands:

```
iface.mapCanvas().clearCache()
iface.mapCanvas().refresh()
```

The symbology of a modified `QgsVectorLayer` instance saved in the `myVector` variable can be updated with the following code:

```
iface.layerTreeView().refreshLayerSymbology(myVector.id())
```

Editing through QgsVectorDataProvider

Each `QgsMapLayer`, as a `QgsVectorLayer` instance, has its own data provider, which can be obtained with the `dataProvider()` method. This is shown in the following code snippet, which is executed in the Python console:

```
myVector =
 QgsVectorLayer("/qgis_sample_data/shapfiles/alaska.shp",
 "MyFirstVector", "ogr")
QgsProject.instance().addMapLayer(myVector)
myDataProvider = myVector.dataProvider()
print(myDataProvider)
```

This will print something similar to `<qgis._core.QgsVectorDataProvider object at 0x000000000649AAF8>` on the Python console. This is the instance of the `QgsVectorDataProvider` class.

The data provider will directly access the stored data, avoiding any control by QGIS. This means that no undo and redo options will be available, and no events related to the editing actions will be triggered.

The code snippets that follow describe how to interact with vector data directly using the data provider. These code samples will modify the Alaska shapefile set of files, so it's better to have a copy of the original files to restore them after you apply the following examples.

Some of the following examples will work directly on features of `myVector`. We can get a feature directly using `QgsFeatureRequest`. For example, to get the feature with the ID 599, we can use the following code snippet:

```
features = myVector.getFeatures(QgsFeatureRequest(599))
myFeature = next(features)
```

Feature 599 is the biggest polygon available in the Alaska shapefile. Remember that `getFeatures` returns an iterator and not the feature directly, so it's necessary to use `next()` to get it.

The `next` function could generate a `StopIteration` exception when the iterator arrives at the end of the features list; for example, when the `getFeatures` result is empty. In our case, we don't care about this exception because we are sure that the feature with the ID 599 exists.

Changing a feature's geometry

After getting feature 599, we can change its geometry. We'll substitute its current geometry with its bounding box. The code snippet to do this is as follows:

```
oldGeom = myFeature.geometry()
bbox = oldGeom.boundingBox()
newGeom = QgsGeometry.fromRect(bbox)
newGeomMap = {myFeature.id() : newGeom}
myDataProvider.changeGeometryValues(newGeomMap)
```

After we refresh the canvas, the shapefile will appear as in the following screenshot:

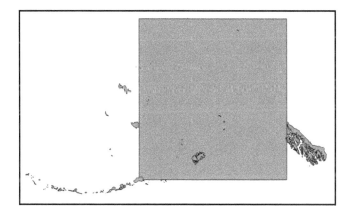

The `changeGeometryValues()` method accepts a `QgsGeometryMap` object. This is the Python dictionary that contains the ID of the feature to change as a key and the new geometry as a value.

Changing a feature's attributes

After getting feature 599, we can change its attributes in a way similar to how we changed the geometry of the feature in the previous subsection.

In this case, the map is a `QgsChangedAttributesMap` class that will be a Python dictionary. This is composed of a key, the ID of the feature to change, and another dictionary as a value. This last dictionary will have the index of the column to change as the key and the new value that has to be set as the value. For example, the following code snippet will change the area value to 0 for column 2 of feature 599:

```
columnIndex = myVector.fields().indexFromName("AREA_MI"))
newColumnValueMap = {columnIndex : 0 }
newAttributesValuesMap = {myFeature.id() : newColumnValueMap}
```

```
myDataProvider.changeAttributeValues(newAttributesValuesMap)
```

You can check whether the value of the parameter has changed by navigating to **View** | **Identity Features**.

If you want to change more than one attribute, you just have to add more key/value pairs in the newColumnValueMap dictionary, using the following syntax:

```
newColumnValueMap = { columnIndex1:newValue1, ...,
  columnIndexN:newValueN}
```

Deleting a feature

A feature can be deleted by pointing at it with its ID. The ID of the feature can be obtained with the id() method of the QgsFeature class. The following snippet will remove feature 599:

```
myDataProvider.deleteFeatures([599])
```

After you refresh the canvas, myVector will be shown as in the following screenshot:

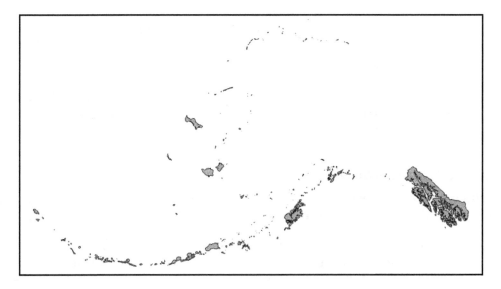

Adding a feature

After we reload the original Alaska shapefile, we will again get feature 599, which we will use as a base to create a new feature. The geometry of this new feature will be set as the bounding box of feature 599; this is done using the following code:

```
# get data provider
myDataProvider = myVector.dataProvider()
# get feature with id 599
features = myVector.getFeatures(QgsFeatureRequest(599))
myFeature = next(features)
# create geometry from its bounding box
bbox = myFeature.geometry().boundingBox()
newGeom = QgsGeometry.fromRect(bbox)
# create a new feature
newFeature = QgsFeature()
# set the fields of the feature as from myVector
# this step only sets the column characteristic of the feature
# not its values
newFeature.setFields( myVector.fields() )
# set attributes values
newAttributes = [1000, "Alaska", 2]
newFeature.setAttributes(newAttributes)
# set the geometry of the feature
newFeature.setGeometry( newGeom )
# add new feature in myVector using provider
myDataProvider.addFeatures([newFeature])
```

The preceding code is explained in the inline comments, and it adds a new feature. The new feature can be checked by opening the attribute table and selecting the last record. This will produce an interface that is similar to the following screenshot:

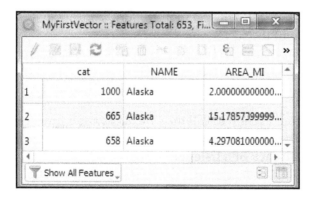

When you select the last record in the attribute table, it will be highlighted.

Editing using QgsVectorLayer

Editing using `QgsVectorLayer` gives you much more power to interact with the QGIS interface and to control the editing flow.

The `QgsVectorLayer` methods to modify the attributes or geometry of a feature are slightly different than the method used in `QgsVectorDataProvider`, but the main characteristic is that all these methods work only if the layer is in the editing mode, otherwise they return `False` to notify failure.

An editing session on the `myVector` vector follows the steps described in the following pseudocode:

```
myVector.startEditing()
< do vector modifications saved in myVector.editBuffer() >
if <all ok>:
myVector.commitChanges()
else:
myVector.rollback()
```

Each step generates events that can be cached if they are useful for our processing scopes. The list of generated events can be read in the `QgsVectorLayer` documentation. For example, `startEditing()` will generate the `editingStarted()` event, adding a feature will generate the `featureAdded()` event, committing changes will emit the `beforeCommitChanges()` event before applying changes, and then the `editingStopped()` event will be generated.

A useful exercise for you is to create a simple script to connect all these events to `print` commands. This is a good way to learn the event sequence that is generated during vector editing.

For example, the following code snippet in the editing console will print a message every time someone starts to edit the `myVector` vector layer:

```
def printMessage():
    print "Editing is Started"
    myVector.editingStarted.connect(printMessage)
```

In the last line, we instruct Python to call the `printMessage` function every time the `editingStarted` event is emitted by `myVector`. In this case, the `printMessage` function is usually known as a **callback** or a **listener**.

Discovering the QgsVectorLayerEditBuffer class

It's possible to have fine-grained control of the editing session managing event of the `QgsVectorLayerEditBuffer` class, which stores all modifications of the layer. It's also possible to access the buffer using the vector layer with the following code:

```
myEditBuffer = myVector.editBuffer()
```

A detailed description of this class is outside the scope of this chapter, but it's strongly suggested to explore it to discover all the PyQGIS editing opportunities, such as setting the attribute of a new feature based on some parameters of the nearest geometry of another layer.

Changing a feature's geometry

After we have reloaded the original Alaska shapefile and got feature 599, we can change its geometry. We'll substitute the current geometry with its bounding box, and the code snippet to do this is as follows:

```
oldGeom = myFeature.geometry()
bbox = oldGeom.boundingBox()
newGeom = QgsGeometry.fromRect(bbox)
myVector.startEditing()
myVector.changeGeometry(myFeature.id(), newGeom)
myVector.commitChanges()
```

In this case, it's not necessary to refresh the canvas because a canvas refresh is triggered by events generated during a commit. After the commit, the interface will appear similar to the screenshot shown in the *Adding a feature* section.

In the preceding code, you can see that the changing geometry has a different API using the data provider. In this case, we can change a feature each time.

Changing a feature's attributes

After we get feature 599, we can change its attributes in a way similar to how we changed the geometry of the feature in the previous section. The following snippet will change the area value to 0 for column 2 of feature 599:

```
columnIndex = myVector.fields().indexFromName("AREA_MI")
myVector.startEditing()
myVector.changeAttributeValue(myFeature.id(), columnIndex, 0)
myVector.commitChanges()
```

You can check whether the value of the parameter has been changed by navigating to **View | Identity Features**.

In the preceding code, notice that the changing attribute has a different API using the data provider. In this case, we can change features one at a time.

Adding and removing a feature

Since the procedure is really similar to those applied to using the data provider, you can test these actions by removing and adding features inside an editing session.

Running Processing Toolbox algorithms

QGIS's versatility is due mainly to two reasons. The first is the ability to customize it by adding functions, thanks to its plugin structure. The second is the power of the Processing Toolbox, which can connect different backend algorithms such as R, GRASS GIS, SAGA, GDAL/OGR, Orfeo Toolbox, OSM Overpass, and many more with dedicated providers.

In this way, for example, we can access all GRASS processing algorithms by using QGIS as the project and presentation manager. Another important ability of the Processing Toolbox is that it can be used to join together all the backend algorithms, allowing you to connect the best algorithms. For example, we can connect GRASS as a producer for another algorithm that is better developed in another backend, such as SAGA. Here, QGIS processing becomes the place where you can add your specific algorithm in a more complex and integrated workflow.

This section is focused on how to code the execution of algorithms that are already available in the Processing toolbox. The main points to learn are as follows:

- Looking for a Processing Toolbox algorithm
- Discovering parameters accepted by the algorithm
- Running the algorithm

In the following sections, we will be using the processing commands after we have imported the processing module with the following code:

```
import processing
```

If the processing module is not imported, every processing command will generate an error such as the following one:

```
Traceback (most recent call last):
File "<input>", line 1, in <module>
NameError: name 'processing' is not defined
```

Listing all available algorithms

Using `processingRegistry()`, you can get information on processing providers, algorithms, parameters, and outputs. Using the `algorithms()` method, we get a list, and using a loop, you can list all the useful information. We can use the following code to get the name of all the algorithm IDs and their corresponding names:

```
for algo in QgsApplication.processingRegistry().algorithms():
    print(algo.id(), "------", algo.displayName())
```

Here, `id()` gives us both the provider name and algorithm name, and `displayName()` gives us a short descriptive name. This returns all the available algorithms in processing:

```
3d:tessellate ------ Tessellate
gdal:aspect ------ Aspect
gdal:assignprojection ------ Assign projection
gdal:buffervectors ------ Buffer vectors
gdal:buildvirtualraster ------ Build virtual raster
gdal:cliprasterbyextent ------ Clip raster by extent
gdal:cliprasterbymasklayer ------ Clip raster by mask layer
gdal:clipvectorbyextent ------ Clip vector by extent
gdal:clipvectorbypolygon ------ Clip vector by mask layer
gdal:colorrelief ------ Color relief
gdal:contour ------ Contour
gdal:convertformat ------ Convert format
gdal:dissolve ------ Dissolve
```

Getting algorithm information

Every processing algorithm has its own GUI with input and output parameters. You may wonder about the parameters, their names, and the values that may be used in the same algorithm using PyQGIS scripts. To discover these elements, we will use the `algorithmHelp` processing command.

This command accepts the command name that is used internally by `processing` as a parameter; for example, the `gdal:cliprasterbyextent` string. We can get help for this command by using the following code:

```
import processing
processing.algorithmHelp("gdal:cliprasterbyextent")
```

The preceding code snippet will produce the following output, which would depend on the installed version of GDAL:

```
Clip raster by extent (gdal:cliprasterbyextent)
----------------
Input parameters
----------------
INPUT: Input layer
Parameter type: QgsProcessingParameterRasterLayer
Accepted data types:
- str: layer ID
- str: ............
PROJWIN: Clipping extent
Parameter type: QgsProcessingParameterExtent
Accepted data types:
- str: as comma delimited list of x min, x max, y min, y max. E.g.
'4,10,101,105'
- str: layer ID. Extent of layer is used.
- str: layer name. Extent of layer is used.
- str: layer source. Extent of layer is used.
.....................................
NODATA: Assign a specified nodata value to output bands
Parameter type: QgsProcessingParameterNumber
....................
OPTIONS: .................
DATA_TYPE: Output data type
Parameter type: QgsProcessingParameterEnum
Available values:
........................
OUTPUT: Clipped (extent)
Parameter type: QgsProcessingParameterRasterDestination
..............................
```

This gives us information about the range of inputs and their data type, along with different values to be used. It also gives us information on the data type of the output. Using this information, we can run different processing algorithms.

You can get more information on algorithms for Processing Toolbox from this site: `https://docs.qgis.org/testing/en/docs/user_manual/processing/console.html`.

Running algorithms from the console

Running a processing algorithm can be done by using the `run` method or the `runandload` method. The first one generates the output without visualizing it, and is useful when you want to generate temporary data. The second method loads and visualizes the output layer in QGIS.

The parameters accepted in this method are the processing command name strings followed by all the parameters, depending on the command used. The `run` method takes the `id` of the algorithm as the first argument, and for the second argument takes all the parameters of the function as a dictionary. In this dictionary, the key is the parameter and the value is the specified value of the parameter. This looks something like this:

```
processing.run("algorithm_id", {input_parameter1: value1, input_parameter2:
value2, ...})
```

We will perform an exercise and clip a raster file according to an extent provided by us. To do this, we will use the `elevation.tif` raster file and use an extent to crop this raster. We first load this raster file and visualize it, and then print its extent:

```
myRaster = QgsRasterLayer("F:/Mastering Geospatial Development with QGIS
3.2/Code/Code/Chapter 9/elevation.tif", "MyFirstRaster")
QgsProject.instance().addMapLayer(myRaster)
ext = myRaster.extent()
print("x-minimum: ", ext.xMinimum(), "x-maximum: ", ext.xMaximum(), "\n",
"y-minimum", ext.yMinimum(), "y-maximum", ext.yMaximum())
```

The output of the `print()` will look as follows:

```
x-minimum: 420909.9284966327 x-maximum: 475709.9284966327
y-minimum 4013968.3650440364 y-maximum 4051068.3650440364
```

Now, we want to clip the raster between `440000` and `460000` for x range and between `4030000` and `4045000`; we can provide these values for `PROJWIN` parameters, as we saw in the previous section in the help documents, using `processingAlgorithmHelp("gdal:cliprasterbyextent")`. Looking at the help file, we provide values for other parameters accordingly.

Now, write the following code to get the desired result of clipping to the specified extent:

```
clipped_raster = processing.run("gdal:cliprasterbyextent", {'INPUT':
'F:/Mastering Geospatial Development with QGIS 3.2/Chapter
11/Data/elevation.tif', 'PROJWIN': '440000, 460000, 4030000, 4045000',
'NODATA': 999, 'DATA_TYPE': 0, 'OUTPUT': 'F:/Mastering Geospatial
Development with QGIS 3.2/ Chapter 11/Data/clipped.tif'})
```

In the preceding code, the fourth parameter is the output filename, but it could be set to
`None`. In this case, the output will be generated in a temporary file.

The return value of `run` is also a dictionary with a key and the output name, and the value
of the key is the reference of the filename generated by the algorithm. In the preceding
example, the output dictionary is something like the `{'OUTPUT': '.../Chapter
11/Data/clipped.tif'}` dictionary.

The last step is displaying the layer which we can do by loading a raster layer as usual:

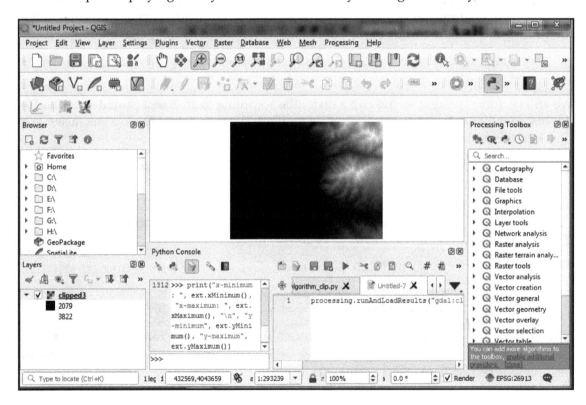

Running your own processing script

Running a custom processing script is not much different from running a generic processing command. We only need to find out how the custom script is addressed by the toolbox.

To discover all this information, we will create and run a simple processing script.

Creating a test Processing Toolbox script

We will start by opening the Processing Toolbox, by navigating to **Processing** | **Toolbox**. Next, navigate to **Scripts** | **Tools** | **Create new script** in the toolbox. These actions will open a new interface, where we can paste the following code snippet:

```
import time
for index in range(100):
    print(index)
    time.sleep(0.1)
```

We will save it with the name `emptyloop`. The default directory where processing will look for user scripts is `<your home dir>/AppData/Roaming/QGIS/QGIS3/profiles/default/processing/scripts/`.

So, our file script will be available with the name `<your home dir>/AppData/Roaming/QGIS/QGIS3/profiles/default/processing/scripts/emptyloop.py`.

The preceding script will count from 0 to 99. The last instruction will wait for `0.1` seconds before running a loop again.

Running the script

We can run the custom script by clicking on the Play button in the **Processing Script Editor**.

This will increment the progress bar and print the progress in the console, as shown in the following screenshot:

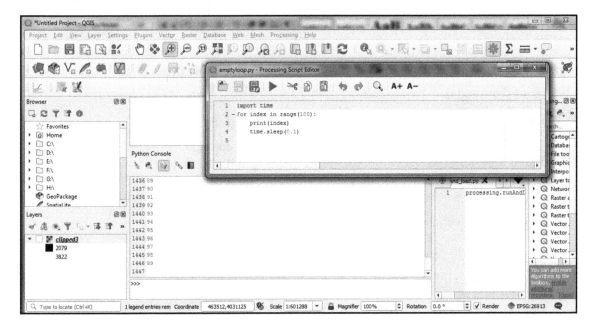

Interacting with the map canvas

A plugin will interact with the map canvas to get some useful information. This information could be, for example, point coordinates or features identified by those coordinates. We can use them to draw geometry entities such as points, lines, or polygons.

Getting the map canvas

The `QgsMapCanvas` class is the class that represents a QGIS canvas. There can be different canvas instances, but the main canvas instance can be referenced with the following code snippet:

```
mapCanvas = iface.mapCanvas()
```

The `QgsMapCanvas` class generates some useful events to support location-based plugins. For example, `xyCoordinates()` sends point locations based on canvas coordinates, and the `keyPressed()` event allows us to know which mouse button has been clicked on the canvas.

Explaining Map Tools

The most powerful method to interact with a map canvas class is to set one of the predefined Map Tools or create a custom one that is derived from the predefined Map Tools classes.

In this chapter, we will not create custom Map Tools using inheritance and overloading, which are two basic concepts of the object-oriented programming paradigm. The following paragraphs are focused on using Map Tools that already exist and customizing them without deriving classes and overloading methods.

The base class for Map Tools is `QgsMapTool`, which has a set of specializations that can be useful for most of the user interaction with the canvas. These derived tools are listed as follows:

- `QgsMapToolEdit` and `QgsMapToolAdvancedDigitizing`: These are focused on managing advanced editing in map coordinates
- `QgsMapToolEmitPoint`: This is focused on intercepting point clicks on the canvas and returning the map coordinates
- `QgsMapToolIdentify`: This is focused on getting layer values at specified point-clicked coordinates
- `QgsMapToolIdentifyFeature`: This is similar to the previous tool, but results also reference the pointed features
- `QgsMapToolPan`: This is focused on managing panning and its events
- `QgsMapToolTouch`: This is focused on managing touch events on the canvas
- `QgsMapToolZoom`: This is focused on managing zoom events

Setting the current Map Tool

For each map canvas, only one Map Tool runs at a time. When it's necessary to set a Map Tool, it's good practice to get the previous one and set it back again when you've finished using the new Map Tool. The following code snippet shows you how to get an old Map Tool, set a new one, and restore the previous one:

```
# import the map tool to use
from qgis.gui import QgsMapToolZoom
# get previous map tool and print it
oldMapTool = iface.mapCanvas().mapTool()
print("Previous map tool is a", oldMapTool)
# create a zoom map tool pointing to the current canvas
# the boolean parameter is False to zoom in and True to zoom out
newMapTool = QgsMapToolZoom(iface.mapCanvas(), False)
# set the current map tool and print it
iface.mapCanvas().setMapTool(newMapTool)
print("Current map tool is a", iface.mapCanvas().mapTool())
#
# here is your code
#
# set the previous map tool and print it
iface.mapCanvas().setMapTool(oldMapTool)
print("Current map tool is a ", iface.mapCanvas().mapTool())
```

The preceding code will generate an output that is similar to the following console lines:

```
Previous map tool is a <qgis._gui.QgsMapToolPan object at
0x0000000055CC70D8>
Current map tool is a <qgis._gui.QgsMapToolZoom object at
0x0000000011FBEA68>
Current map tool is a <qgis._gui.QgsMapToolPan object at
0x0000000055CC70D8>
```

You will notice that changing the current Map Tool with setMapTool() will also change the cursor icon.

Getting point-click values

In this paragraph, we will create code that will be useful for getting point-click coordinates and printing them in the console. We will use the QgsMapToolEmitPoint Map Tool, but the structure of the following code can be applied to other available Map Tools as well.

We will write all the code in the Python console, but it could be used in a custom plugin, for example, to create a GUI interface to trace mouse movement or to plot a polygon that is based on clicked points.

We will take the following steps to get the clicked points:

1. Save the previous Map Tool
2. Create a `QgsMapToolEmitPoint` Map Tool
3. Create an event handler for the map canvas to trace mouse movement
4. Register the event above the event handler to the `xyCoordinate` event
5. Create an event handler for the Map Tool to trace clicked points by setting the following conditions:

    - If the left button is clicked, then print coordinates
    - If the right button is clicked, then restore the previous Map Tool

6. Register the event handler to the click event generated by the Map Tool
7. Activate the new `QgsMapToolEmitPoint` Map Tool

These steps are coded in the next paragraphs.

Getting the current Map Tool

To return to the current Map Tool after setting the new one, we need to save it in a variable that can be used later. We can get the current Map Tool by using the following code snippet:

```
previousMapTool = iface.mapCanvas().mapTool()
```

Creating a new Map Tool

We can create a Map Tool with this simple code snippet:

```
from qgis.gui import QgsMapToolEmitPoint
myMapTool = QgsMapToolEmitPoint(iface.mapCanvas())
```

You will notice that each Map Tool constructor needs a parameter that is the canvas on which it will operate.

Creating a map canvas event handler

An event handler is useful to execute actions that are based on user interaction with the canvas. First, we will create a handler that prints coordinates based on mouse movements on the canvas.

The event handler will receive the parameters passed by the event to which it is attached. The event handler will be attached to the `QgsMapCanvas` event, `xyCoordinates(QgsPoint )`, and it can be coded using the following code snippet:

```
def showCoordinates(currentPos):
    print("move coordinate %d - %d" % (currentPos .x(),
    currentPos.y()))
```

After creating the handler, we have to attach it to the canvas event with the following code:

```
iface.mapCanvas().xyCoordinates.connect(showCoordinates)
```

If you want to remove the handler, use the following code:

```
iface.mapCanvas().xyCoordinates.disconnect(showCoordinates)
```

Ensure that you write the function name as `showCoordinates`, because if you don't pass any parameters to the `disconnect` call, then all handlers attached to the `xyCoordinates` event will be removed.

Creating a Map Tool event handler

An event handler will be attached to the `QgsMapToolEmitPoint` event, `canvasClicked(QgsPoint, Qt.MouseButton)`, and then it can be coded using the following code snippet:

```
# import the Qt module that contain mouse button definitions like
Qt.LeftButton or Qt.RightButton used later
from PyQt5.QtCore import Qt
# create handler
def manageClick(currentPos, clickedButton):
    if clickedButton == Qt.LeftButton:
        print("Clicked on %d - %d" % (currentPos.x(),
        currentPos.y())
    if clickedButton == Qt.RightButton:
        # reset to the previous mapTool
        iface.mapCanvas().setMapTool(previousMapTool)
        # clean remove myMapTool and relative handlers
        myMapTool.deleteLater()
```

After creating the handler, we have to attach it to the Map Tool event using the following code:

```
myMapTool.canvasClicked.connect(manageClick)
```

Setting up the new Map Tool

Now, it's time to activate the new Map Tool to pass canvas control to its event handlers. This can be done with the following code:

```
iface.mapCanvas().setMapTool(myMapTool)
```

After executing the preceding command, the new Map Tool will be activated, and it will print coordinates in the console, as shown in the following screenshot:

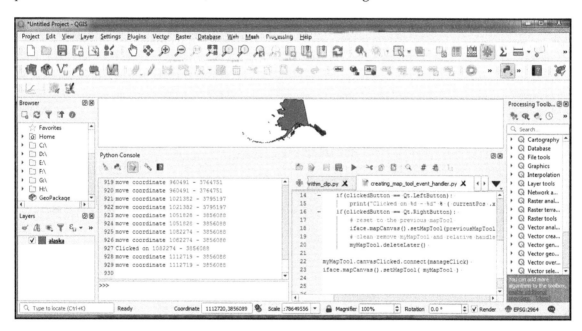

Right-click on the canvas to return to the previous Map Tool, but you still need to remove the canvas event handler. As explained in the previous paragraph, removing the xyCoordinates handler is done by executing the following code:

```
iface.mapCanvas().xyCoordinates.disconnect(showCoordinates)
```

Using point-click values

The previous paragraph explained how to get a point-click coordinate. This coordinate can be used to get the correspondent value in a raster layer or to get the underlying feature of a vector layer.

To identify a feature for a vector layer, it's better to use the dedicated Map Tool, `QgsMapToolIdentifyFeature`, directly, but for a raster layer that has point coordinates, we can use the `QgsRasterDataProvider.identify()` method to get raster values at a specified point.

By loading and selecting the `landcover` raster from `qgis_sample_data/raster`, we can modify the `manageClick` method of the previous example in the following way:

```
def manageClick( currentPos, clickedButton ):
    if clickedButton == Qt.LeftButton:
        provider = iface.activeLayer().dataProvider()
        result = provider.identify(currentPos,
            QgsRaster.IdentifyFormatValue)
        if result.isValid():
            print "Value at %d - %d" % (currentPos.x(), currentPos.y())
            print result.results()
    if clickedButton == Qt.RightButton:
        # reset to the previous mapTool
        iface.mapCanvas().setMapTool(previousMapTool)
        # clean remove myMapTool and relative handlers
        myMapTool.deleteLater()
```

Running the code of the previous paragraph with the `manageClick` function will generate an output similar to the following screenshot:

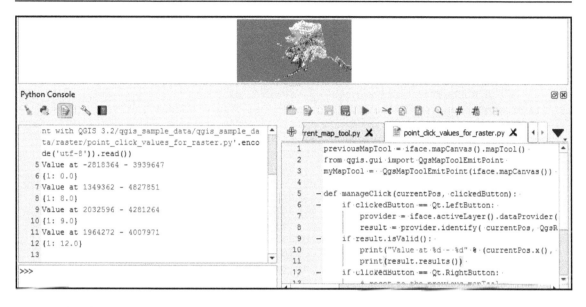

The result of the `identify()` call is a `QgsRasterIdentifyResult` object that contains all the information of the result. In this case, the return of `results()` is a Python dictionary with the band number as the key and the raster value in the clicked point as the value.

Exploring the QgsRubberBand class

A rubber band is a graphical canvas item that can be used to draw geometry elements on the canvas, for example, points, lines, or polygons. It is generally used in combination with a customized Map Tool to get click coordinates that are used to add or move points of the rubber band object.

The following code snippet will upgrade the previous example to get canvas-click coordinates, to draw a polygon in the canvas instead:

```
from PyQt5.QtCore import Qt
from PyQt5.QtGui import QColor
from qgis.core import Qgis
from qgis.gui import QgsMapToolEmitPoint, QgsRubberBand
previousMapTool = iface.mapCanvas().mapTool()
myMapTool = QgsMapToolEmitPoint(iface.mapCanvas())
# create the polygon rubber band associated to the current canvas
myRubberBand = QgsRubberBand(iface.mapCanvas(),
QgsWkbTypes.PolygonGeometry)
# set rubber band style
color = QColor("red")
```

```
color.setAlpha(50)
myRubberBand.setColor(color)
def showCoordinates(currentPos):
    if myRubberBand and myRubberBand.numberOfVertices():
        myRubberBand.removeLastPoint()
        myRubberBand.addPoint(currentPos)
iface.mapCanvas().xyCoordinates.connect(showCoordinates)

def manageClick(currentPos, clickedButton):
    if clickedButton == Qt.LeftButton:
        myRubberBand.addPoint(currentPos)
    # terminate rubber band editing session
    if clickedButton == Qt.RightButton:
        # remove showCoordinates map canvas callback
        iface.mapCanvas().xyCoordinates.disconnect(showCoordinates)
        # reset to the previous mapTool
        iface.mapCanvas().setMapTool(previousMapTool)
        # clean remove myMapTool and relative handlers
        myMapTool.deleteLater()
        # remove the rubber band from the canvas
        iface.mapCanvas().scene().removeItem(myRubberBand)
myMapTool.canvasClicked.connect( manageClick )
iface.mapCanvas().setMapTool(myMapTool)
```

Executing the preceding code and clicking on the canvas with the left mouse button will produce results similar to the following screenshot:

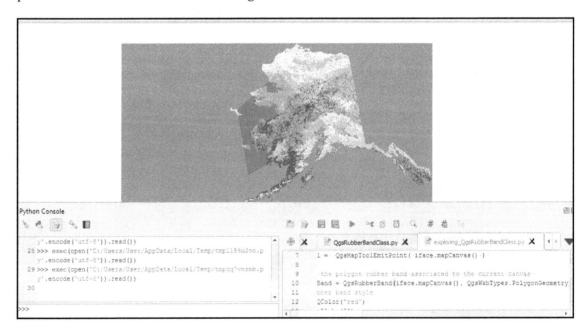

The image in the preceding screenshot has the `alaska.shp` layer as a base map to show the evidence of the transparency of the rubber band drawn.

If you see an error showing `Qgis not found` or `ImportError: No module named Qgis.core`, you need to set your `PATH` and `PYTHONPATH` by setting them according to your OS, as shown here:

```
OSGeo4W Shell
run o-help for a list of available commands
C:\>set PATH="C:\Program Files\QGIS 3.6\bin"

C:\>set PATH="C:\Program Files\QGIS 3.6\apps\qgis\bin"

C:\>set PYTHONPATH="C:\Program Files\QGIS 3.6\apps\qgis\python"

C:\>set PYTHONPATH="C:\Program Files\QGIS 3.6\apps\qgis\python\plugins"

C:\>set PYTHONPATH="C:\Program Files\QGIS 3.6\apps\Python37\Lib\site-packages"

C:\>
```

Summary

This chapter offered you a simplified way to interact with QGIS, and was more oriented toward a GIS analyst than a GIS programmer or a computer scientist. The chapter was also oriented to GIS companies that are interested in reusing code that has already been developed, but was probably developed for GIS platforms that are not free.

This chapter provided basic knowledge about how to interact with QGIS using the PyQGIS programming language.

We looked at how to programmatically load different kinds of layers, from raster to vector. We also explained vectors and how to manage different kinds of vector resources, from filesystems to remote database connections. We also explored the vector structure in more detail, and you learned how to browse and edit its records. Different kinds of editing workflows were proposed in the chapter so that you can interact with the QGIS framework in a better manner.

You learned how to launch Processing Toolbox algorithms and user-developed Processing Toolbox scripts to enhance QGIS with new functionalities. You also learned how to launch external commands or scripts to offer a way to integrate previously developed code, and thereby reduce development and testing costs.

Lastly, we explored the QGIS map canvas and how to interact with it using Map Tools. We obtained canvas coordinates and created new Map Tools to draw canvas objects. This has given you the basic skills needed to create new plugins that interact with layers displayed on the QGIS canvas.

Other Books You May Enjoy

If you enjoyed this book, you may be interested in these other books by Packt:

Learn QGIS - Fourth Edition
Andrew Cutts, Anita Graser

ISBN: 9781788997423

- Explore various ways to load data into QGIS
- Understand how to style data and present it in a map
- Create maps and explore ways to expand them
- Get acquainted with the new processing toolbox in QGIS 3.4
- Manipulate your geospatial data and gain quality insights
- Understand how to customize QGIS 3.4
- Work with QGIS 3.4 in 3D

Hands-On Geospatial Analysis with R and QGIS
Shammunul Islam

ISBN: 9781788991674

- Install R and QGIS
- Get familiar with the basics of R programming and QGIS
- Visualize quantitative and qualitative data to create maps
- Find out the basics of raster data and how to use them in R and QGIS
- Perform geoprocessing tasks and automate them using the graphical modeler of QGIS
- Apply different machine learning algorithms on satellite data for landslide susceptibility mapping and prediction

Leave a review - let other readers know what you think

Please share your thoughts on this book with others by leaving a review on the site that you bought it from. If you purchased the book from Amazon, please leave us an honest review on this book's Amazon page. This is vital so that other potential readers can see and use your unbiased opinion to make purchasing decisions, we can understand what our customers think about our products, and our authors can see your feedback on the title that they have worked with Packt to create. It will only take a few minutes of your time, but is valuable to other potential customers, our authors, and Packt. Thank you!

Index

2

2.5D renderer
  using 244, 246, 247, 248
2.5D vector
  styling 88, 89, 90

3

3D views
  creating 249

A

address geocoding
  about 188
  with local street network data 192, 194, 195
  with web services 190, 191, 192
  working 189, 190
advanced field calculations
  about 149
  conditionals 153, 154
  conditions 153
  current date, calculating 151
  current date, formatting 151
  operators 153
  with geometry 152
  writing 151
algorithms
  adding, to model 335, 336, 338
analysis
  performing, SAGA used 295
Android
  QGIS, installing 12
API documentation 359
Application Programming Interfaces (APIs) 358
Aptana
  installing 384, 386, 388
area mask

analyzing 242, 244
Atlas configuration 250, 252
Atlas
  coverage feature, highlighting 255, 256
  creating 250
  dynamic legends, generating 255
  dynamic titles, generating 252, 254
attribute table
  geometry columns, adding to 135

B

batch processing
  using, with models 354
blending modes 32
Browser panel 116
Buffer tool 142
buffers
  creating 142, 143, 144

C

CAD-style digitizing tools
  about 174
  circle, adding 174
  circle, adding from three points 176
  circle, adding from two points 175, 176
  rectangle, adding 177
  rectangle, adding from Extent 178
  rectangle, adding from its center point 178
  rectangle, adding from three points 178
  regular polygon, adding 178
  regular polygon, adding from center 179
  regular polygon, adding from corner 179
  regular polygon, adding from two points 179
canvas
  updating 413
categorized vector
  styling 76, 77

changeable panels, color picker
  about 43
  Color ramp 43
  Color sampler 46
  Color swatches 44, 46
  Color wheel 44
Chromebook
  QGIS, installing 13
Clip grid with polygon
  used, for clipping land 298
Clip tool 136, 138
coastal vignettes
  creating 240, 242
color picker components
  about 42, 43
  in Layers panel 47, 48, 49
color ramps
  adding 53
  editing 59
  exporting 51
  importing 52, 53
  managing 49
  paletted raster band rendering 59, 61
  removing 51
  renaming 51
  singleband gray raster band rendering 61, 63,
    64
  singleband pseudocolor raster band rendering
    64, 66
  singleband rasters, styling 59
ColorBrewer color ramp
  adding 56
colors
  managing 39, 40, 41
  selecting 39, 40, 41
control feature rendering order 93
Convex Hull tool 145
convex hulls
  generating 145
cookie cutter 136
coordinate reference system (CRS)
  about 9, 145, 196
  parameters 146
  working with 22
coverage feature

highlighting 255, 256
cpt-city color ramp
  adding 57, 58
current date
  calculating 151
  formatting 151
Custom Coordinate Reference System Definition
    (Custom CRS) tool 145
custom coordinate reference system
  defining 147
  Definition window 148
custom QGIS variables 35

D

data inputs
  adding, into model 332, 333, 335
Data Plotly plugin
  working with 262, 263, 265
data
  editing, into QGIS Desktop 26
  importing, in GeoPackage 112
database tables 108
database vectors
  managing 404, 406
database
  about 108
  loading, into QGIS Desktop 21
debugger 384
debugging environment
  setting up 383
default styles
  adding, in QGIS project 105
  loading 103, 104
  renaming, in QGIS project 105
  restoring 104
  saving 103, 104
  setting 103, 104
Delaunay triangulation 130
Densify geometries tool 133
diagram types
  parameters 95
diagrams
  use, to display thematic data 94
Difference tool 136, 139, 140
Digital Elevation Model (DEM) 278

Digital Terrain Model (DTM) 278
Dissolve tool 145
dynamic legends
  generating 255
dynamic titles
  generating 252, 254

E

editing features
  about 174
  CAD-style digitizing tools 174
  vertex tool 179, 180

F

feature blending mode 92
feature's attributes
  changing 415, 419
feature's geometry
  changing 415
features
  adding 417, 420
  canvas, updating 413
  deleting 416
  densifying 133
  dissolving 145
  editing 413
  editing, through QgsVectorDataProvider 414
  editing, with QgsVectorLayer 418
  iterating over 411, 412
  removing 420
  simplifying 133
  symbology, updating 413
field calculator interface
  exploring 149
field calculator window
  sections 149
file structure, plugin_first
  main plugin files, exploring 375
FOSSGIS packages
  installing, on Ubuntu Linux 12
Free and Open Source Software for Geographical
    Information Systems (FOSSGIS) 11
functionality
  adding, with plugins 33, 34
functions

types 150

G

GDAL Proximity
  used, for surface water proximity 299, 301
GDAL Raster calculator
  used, for querying water proximity 301, 302
Geodetic Parameter Dataset 22
Geographic Information System (GIS) 121
Geographical Resources Analysis Support System
    (GRASS) library
  reference 284
Geographical Resources Analysis Support System
    (GRASS)
  about 10, 269
  least-cost path, calculating 280
  raster analysis, performing with 276
  shaded relief, calculating 270, 279
  viewshed, evaluating 288
geometry columns
  adding, to attribute table 135
Geometry Engine Open Source (GEOS) 145, 410
geometry generators
  features 258, 259, 261
GeoPackage
  connecting to 112
  data, importing into 112
  layer, importing from map canvas 113
  table, creating within 114
  table, deleting 116
  tables, exporting from 116
  vector file, importing into 112
  vector layer, importing options 114
Geoprocessing Tools
  Clip tool 136, 138
  Difference tool 136, 139, 140
  Intersect tool 140
  spatial overlay tools 136
  Symmetrical Difference tool 140
  using 135, 136
Georeferencer GDAL plugin
  using 197, 199, 202
georeferencing, with dataset
  about 202, 204
  ground control points, entering 205, 207

operation, completing 213
transformation settings 207, 209, 210, 212
georeferencing
about 196
Georeferencer GDAL plugin, using 197, 199,
202
ground control points (GCP) 196
with point file 214, 216
Geospatial Data Abstraction Library (GDAL) 20,
145
gradient color ramp
adding 53, 55
graduated vector
styling 77, 78, 79
Graphical Modeler (GM)
about 325
opening 326
using 325
ground control points (GCP) 196
GUI logic
modifying 379

H

heatmap vector
styling 86, 87
histogram chart diagram
creating 102, 103
hydrologic analysis, exploring with SAGA
about 306
pits, removing from DEM 306, 308, 311
streams, delineating 316
streams, deriving 312
streams, selecting 313, 315
upstream area, calculating above Fort Klamath
317, 320, 321, 322
hydrologic analysis
exploring, with SAGA 306

I

iface class
exploring 365, 366
Internet Relay Chat (IRC) 360
Intersect tool 140
inverted polygon shapeburst fills
used, for analyzing area mask 242, 244

used, for creating beautiful effects 239
used, for creating coastal vignettes 240, 242
inverted polygons vector
styling 84, 86
iterators 413

L

layer blending mode 91
layer transparency 91
Layers panel
color picker components 47, 48, 49
layers
loading, in QGIS 400
Least-cost path (LCP)
r.cost, used for calculating cumulative cost
raster 285
about 286
calculating 280
land use raster, reclassifying 282, 283
LCP, used for calculating cost path 286, 287
reclassified slope, combining with land use
layers 284
slope raster, reclassifying 282, 283
slope, calculating with r.slope 281
lines
converting, into polygons 125, 126
polygons, converting into 126, 127
live layer effects
using 234, 235, 237, 238
local street network data
used, for geocoding 192, 194, 195
long-term release (LTR) 10

M

macOS
QGIS, installing 11
mailing lists 360
map canvas
interacting with 426
obtaining 426
point-click values, obtaining 428
QgsRubberBand class, exploring 433, 435
Map Tools
explaining 427
setting 428

maps
  composing 32, 33
Massive Open Online Course (MOOC) 399
Merge Vectors tool 122
model algorithms
  executing 346, 348
modeler
  configuring 328, 329, 331
models
  algorithms, adding 335, 336, 338
  batch processing, using 354
  converting, into Python script 356
  data inputs, adding 332, 333, 335
  documenting 344, 345
  editing 342, 343
  exporting 345
  loading 345
  naming 328, 329, 331
  nesting 349, 352, 354
  running 339, 341
  saving 345
  working with 332
multiband rasters
  styling 68
multipart features
  converting, to singleparts features 134

N

normalization 109

O

Open Source Geospatial Foundation (OSGeo) 9
  reference 145

P

paletted raster band rendering 59, 61
parameters, diagram types
  attributes, adding 98
  diagram placement parameters 96, 97
  diagram size parameters 95
pie chart diagram
  creating 99, 100
Plugin Builder
  installing 368
  used, for creating plugin structure 367

plugin example 376
plugin logic
  code detail 382, 383
  combo box, populating 381
  dialog, running 382
  dialog, showing 382
  layers, classifying 380
  modifying 379, 380
plugin structure
  creating, with Plugin Builder 367
plugin_first
  creating 369
  file structure 374
  icon resource, compiling 372, 373, 374
  mandatory plugin parameters, setting 370, 371
  optional plugin parameters, setting 371
  plugin Builder-generated files 376
  plugin code, generating 372
plugins
  locating 368
  used, for adding functionality 33, 34
point cluster vector
  styling 84
point-click values
  current Map Tool, obtaining 429
  map canvas event handler, creating 430
  Map Tool event handler, creating 430
  new Map Tool, creating 429
  new Map Tool, setting up 431
  obtaining 428
  using 432, 433
point-displacement vector
  styling 82, 84
points
  creating, from coordinate data 181, 182, 184, 185
polygon centroids
  creating 124, 125
polygon layers
  overlaying, with Union 141, 142
polygons
  converting, into lines 126, 127
  gap, repairing between 230, 232
  generating 127
  lines, converting into 125, 126

overlap, repairing between 229, 230
Voronoi polygons 127, 128, 129
primary key 109
processing toolbox algorithms
  custom processing script, running 425
  information, obtaining 421, 422
  listing 421
  running 420
  running, from console 423, 424
Processing Toolbox
  about 269, 325
  algorithms, running 274, 275
  configuring 271, 272
  features 270
  using 276
  viewing 272
PROJ.4 definition format 146
PROJ.4 project website
  reference 146
PyDev project
  creating, for TestPlugin 393, 395
Pydevd server
  QGIS, connecting 391
  starting 390, 391
PyQGIS code snippet
  layer features, looping 364
PyQGIS cookbook
  about 358
  reference 358
Python Console
  about 362, 363
  PyQGIS code snippet 364
  QGIS API, exploring 366
  QGIS sample data, obtaining 363
Python
  about 400
  resources 399
PYTHONPATH
  setting up 388, 389

Q

QField 12
QGIS 3.x
  features 13, 14, 15
  tools 13

QGIS API
  exploring, in Python Console 366
QGIS Browser 16
QGis class
  exploring 365, 366
QGIS color ramp collection
  managing 49, 50
QGIS community 359
QGIS Desktop
  about 16, 17
  blending modes 32
  data, editing 26
  data, loading 17
  databases, loading 21
  raster data, loading 19
  raster data, styling 30
  snapping 27
  vector data, loading 18, 19
  vector data, styling 28, 30
  web services, loading 21
QGIS documentation
  reference 358
QGIS processing
  reference 271
QGIS Raster calculator tool 297, 304
QGIS
  about 9
  connecting, to Pydevd server 391
  downloading 10
  installing 10
  installing, on Android 12
  installing, on Chromebook 13
  installing, on macOS 11
  installing, on Ubuntu Linux 11, 12
  installing, on Windows 10
  layers, loading 400
  release schedules 10
QgsGeometry
  exploring 410
QgsRasterLayer class
  exploring 402
QgsRubberBand class
  exploring 433, 435
QgsVectorDataProvider
  features, editing 414

QgsVectorLayer
  used, for editing features 418
QgsVectorLayerEditBuffer class
  discovering 419
  feature's geometry, changing 419
Qt Designer
  two pull-down menus, adding 378
  used, for modifying layout 377
queries
  creating 117

R

r.cost tool 285
r.slope
  used, for calculating slope 281
r.viewshed tool 290
random color ramp
  adding 55
raster color
  rendering 68, 71
raster data
  loading, into QGIS Desktop 19
  styling 30
raster file type
  converting 158
  converting, with Translate tool 158, 159
rasters
  clipping 161, 163
  converting, into vectors 163, 164
  exporting 160
  exporting, to GeoPackage 160, 161
  managing 401
  merging 155, 157
  QgsRasterLayer class, exploring 402
  reclassifying 166, 168
  resampling 71, 72
  visualizing, with QGIS layer manager 403
Reclassify grid values tool
  used, for reclassifying land use 303
Remote Debug QGIS plugin
  used, for connecting PYTHONPATH to PyDev
    server 392
RollApp
  about 13
  URL 13

rows, vector structure
  QgsGeometry, exploring 410
rule-based vector
  styling 79, 81, 82

S

SAGA Raster calculator
  used, for calculating elevation ranges 296, 297
  used, for combining raster layers 304, 305
  used, for querying land 298
select statements
  using 117, 118
shaded relief
  calculating 278
Simplify geometries tool 133
single-symbol vector
  styling 73, 74
singleband gray raster band rendering 61, 63, 64
singleband pseudocolor interpolations 67
singleband pseudocolor raster band rendering 64,
  66
singleband rasters
  styling 59
singleparts features
  converting, to multipart features 134
snapping 27
spatial database
  creating, in QGIS 110, 112
spatial overlay tools 136
spatial view
  creating 118
  dropping 118
SpatiaLite (SQLite) 111
Stack Exchange community
  about 361
  issues, reporting 361, 362
  knowledge, sharing 361, 362
statistical summary
  viewing, of vector layers 148
Structured Query Language (SQL)
  reviewing 109, 110
symbology
  updating 413
Symmetrical Difference tool 140
System for Automated Geoscientific Automation

(SAGA)
about 269
habitat, evaluating 296
hydrologic analysis, exploring with 306
used, for performing analysis 295

T

table joins
creating 24, 25
table relationships 109
table
about 108
column, editing 115
creating, within GeoPackage 114
deleting, in GeoPackage 116
exporting, from GeoPackage 116
field, editing 115
renaming 115
working with 22, 24, 114
test processing toolbox script
creating 425
running 425
TestPlugin
basic logic, adding 376
breakpoints, adding 395
debugging 392, 396
PyDev project, creating for 393, 395
text diagram
creating 101
text representations of geometry
mapping 185, 187, 188
thematic data
displaying, with diagrams 94
Toolbox function 272
topological errors, repairing with topological editing
duplicate geometries, resolving 225
editing parameters, setting 226, 227, 228
overlaps, repairing 225
topological errors
repairing, via topological editing 224
topological rules
about 218
for line features 218
for point features 218
for polygon features 219

topology checker
installing 217
using 220, 222, 224
topology
checking of vector data 217
Translate tool
used, for converting raster file type 158, 159
Triangulated Irregular Networks (TINs) 130

U

Ubuntu Linux
FOSSGIS packages, installing 12
QGIS, installing 11, 12
Uniform Resource Identifier (URI) 404
Union tool
about 141
polygon layers, overlaying 141, 142
US National Geodetic Survey (NGS) 214

V

vector data
loading, into QGIS Desktop 18, 19
styling 28, 30
topology, checking 217
vector files
managing 404
vector geometries
converting 124
vector layer rendering
about 91
control feature rendering order 93
feature blending mode 92
layer blending mode 91
layer transparency 91
vector layers
statistical summary, viewing of 148
vector methods 406, 407
vector structure
about 406
describing 407
header 408
rows 409
vectors
converting, to raster format 164, 166
merging 122, 123

styling 72
Version Control System (VCS) 371
vertex tool 179, 180
vertices
  extracting 132, 133
views
  creating 117
viewshed
  calculating, for towers with r.viewshed 290, 291, 292
  combining, with r.mapcalc.simple 292, 294, 295
  evaluating 288
  GDAL, used for clipping elevation to park
    boundary 289

Voronoi polygons
  using 127, 128, 129

W

web services
  loading, into QGIS Desktop 21
  used, for address geocoding 190, 191, 192
Well-Know Text (WKT) 181, 183, 411
Windows, QGIS installation options
  OSGeo4W Network Installer 11
  QGIS Standalone Installer 10
Windows
  QGIS, installing 10

www.ingramcontent.com/pod-product-compliance
Lightning Source LLC
Chambersburg PA
CBHW060644060326
40690CB00020B/4514